A BIRDWATCHERS' GUIDE TO SOUTHERN SPAIN AND GIBRALTAR

CLIVE FINLAYSON

Illustrations by Mike Langman

BIRD
WATCHERS'
GUIDES

Prion Ltd.
Perry

ACKNOWLEDGEMENTS

I am grateful to my wife, Geraldine, for her support and for drawing the maps. For their companionship in the field while visiting many of the sites mentioned in this guide I am grateful to Paul Acolina, Andrew Fortuna, Steve Holliday, John Licudi, Mario Mosquera, Ernest Pardo, Harry Van-Gils, Alberto Vega, John Cortes, Charles Perez, Darrien Ramos, Nigel Ramos and Jesus Parody.

To Geraldine

Clive Finlayson was born in Gibraltar in 1955. He read Zoology at the University of Liverpool and studied the biology of Swifts and Pallid Swifts for his doctorate at the Edward Grey Institute, University of Oxford. Clive has been interested in birds from an early age when he was captivated by the migration of raptors over Gibraltar. He has published many books, including the recently published 'Birds of the Strait of Gibraltar'. He has travelled widely throughout southern Spain, acquiring an intimate knowledge of its countryside and birds.

CONTENTS

INTRODUCTION

Andalucía is the southernmost region of Spain. With an area of over 98,000 sq km it is larger than the United Kingdom and has the lowest population density in Spain. It is essentially composed of two mountain chains with the valley of the Guadalquivir river in between. In the north, the dark undulating hills of the Sierra Morena run from east to west separating the region from the meseta (tableland) of central Spain. South of the Sierra Morena, the Guadalquivir traces its route from the Sierra de Cazorla in Jaén, west to meet the Atlantic Ocean. Near its mouth, south-west of Sevilla, it forms the large complex of marshes known as the Marismas. South of the Guadalquivir valley lie the mountains of the Penibetic chain, the highest point being the Mulhacén (3482m) in the Sierra Nevada - Spain's highest peak. The chain continues west past the Serranía de Ronda towards the Rock of Gibraltar. Gibraltar itself is a 6 sq km limestone peninsula with a peak of 426m which rises directly from the sea. The Rock, a British Dependent Territory, has been the base for ornithologists travelling across Andalucía since the 18th century. The border with Spain, closed for many years, was re-opened in 1985.

Southern Spain and the region of the Strait of Gibraltar have always attracted the attention of ornithologists. The Reverend John White lived on the Rock of Gibraltar in the 1770's and provided his brother Gilbert (of Selborne fame) with information on bird migration. During the 19th and early 20th centuries, naturalists and explorers such as Abel Chapman, Walter Buck, Howard Irby and Willoughby Verner described the birds and the varied landscapes of these southernmost areas of Europe. In recent years interest in the area has increased, and in the past decade there has been a great influx of visiting birdwatchers.

One of the main ornithological features of southern Iberia and of the Strait of Gibraltar in particular, is bird migration. Many birdwatchers come to the Strait of Gibraltar at peak migration time to see raptors, but because they do not understand how winds modify the flylines, they can spend hours or even days watching from the wrong observation point and seeing very little. In this Guide, a substantial part of the main body of text describes sites where migration of the major groups (raptors, storks, seabirds, waders, waterbirds and passerines) can best be observed. With migrants, it is difficult to be precise as to what can be seen where and when. Circumstances at a site can change even within a day and the species list only includes the most interesting birds to be seen at that particular site. Migration-watching has the thrill of unpredictability with unlikely species suddenly appearing or with large concentrations of birds occurring within a brief period. By watching the migration sites mentioned in the text and following the advice given on winds, birdwatchers should be able to experience the phenomenon of migration at its best.

It is easier to predict where birds can be found when they are breeding or wintering. Southern Iberia has some very interesting habitats and special breeding birds. Among these habitats, the wetlands with some rare waterbirds and the mountains with spectacular raptor colonies stand out. Sites have been chosen to provide opportunities to see the typical nesting birds of the area including the rare ones. However, specific information on raptor nesting sites or on some of the lesser-known lagoons has been omitted. Some species are now very scarce and if detailed information were to fall into the hands of egg collectors or over-zealous photographers it could have a disastrous effect. If visitors come across the nesting site of a raptor or other scarce species they should avoid the temptation of letting others know.

Southern Iberia must be preserved as one of Europe's prime ornithological areas. Andalucía has probably lost Bittern, Lammergeier and Andalucian Hemipode as breeding birds while many other species are now very rare. By encouraging birdwatchers to visit the area and promoting an interest in the region's wildlife, a great deal can be done towards the protection of key sites. All visitors should be conscious of their actions and not cause unnecessary disturbance to birds for the sake of better views or a close-up photograph.

There are few resident birdwatchers in the area and many potentially good but poorly known sites. Visitors who discover worthwhile sites which are not included in this guide, or see unusual birds, should send details to the author– Dr Clive Finlayson, The Gibralter Museum, PO Box 939, Gibraltar.

Spanish
Imperial Eagle

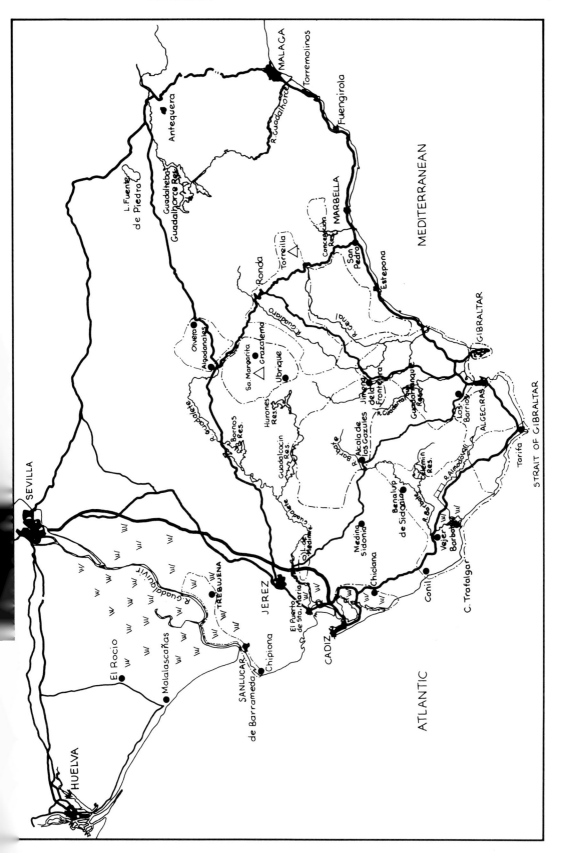

PRE-TOUR INFORMATION

Passports are required for entry into Spain and Gibraltar. European Community Nationals, except citizens of the United Kingdom and Eire who require passports, may enter Spain with a national identity card. EC Nationals, except from Spain and Portugal, may enter Gibraltar with national identity cards. British Subjects (UK or colonies) may use a British visitor's passport, in lieu of a passport, to enter Gibraltar.

Visas are not required for entry into Gibraltar except for nationals of some Arab and communist countries and by stateless persons. Visas are not generally needed for Spain either although they are required of citizens of some Commonwealth countries (e.g. Australia and New Zealand). Visitors are permitted a three month stay.

The consular offices of various countries can be found in the following towns, as indicated below.

Gibraltar: Belgium, Denmark, Finland, France, Greece, Israel, Italy, Netherlands, Norway, Sweden.

Almería: Denmark, Finland, France, Germany, Netherlands.

Cádiz: Belgium, Denmark, Finland, France, Italy, Norway, Netherlands, Sweden.

Huelva: Denmark, Finland, Germany, Netherlands, Norway, Portugal, Sweden.

Málaga: Belgium, Canada, Denmark, Finland, France, Germany, Greece, Iceland, Italy, Netherlands, Norway, Sweden, Switzerland, United Kingdom.

Sevilla: Austria, Belgium, Denmark, Finland, France, Germany, Greece, Italy, Netherlands, Norway, Portugal, Sweden, United Kingdom, United States.

Currency and Exchange Rate

The Spanish unit of currency is the Peseta and many establishments on Gibraltar accept them. The Gibraltar pound is interchangeable with sterling which is accepted everywhere on Gibraltar.

In May 1993 one pound sterling was roughly equivalent to 180 pesestas. Money can be readily exchanged in banks, bureaux de change, some hotels and other establishments.

Customs

The import allowance is the same for Spain and Gibraltar: 200 cigarettes or 100 cigarillos (max. weight each of 3g) or 50 cigars or 250g tobacco; one litre of spirits, liqueurs and cordials or two litres of fortified or sparkling wine or two litres of table wine; perfume 50g and 0.25 litres of toilet water. Personal effects, including binoculars, telescope and camera are permitted.

Health Certificates

No vaccinations are required.

Travelling to Southern Spain and Gibraltar

Spanish Airports

Málaga international airport receives flights from many parts of Europe as well as internal Spanish flights. Its location makes it less convenient for visitors wanting to explore the west of the region but the advantages and disadvantages of each airport must be judged according to individual needs. Sevilla and Jerez also have airports.

Standard facilities at the airports include banks, car hire desks and airline re-confirmation offices. Taxis and buses operate from the airports (see below).

Gibraltar Airport

Gibraltar Airport is regularly used by many passengers travelling to the western parts of the Costa del Sol as well as Gibraltar. It is ideally located for the itineraries described in this book and is serviced by regular flights from London Gatwick and Heathrow with GB Airways. There are flights from London seven days a week and there is also a weekly flight from Manchester airport.

Travelling in Southern Spain and Gibraltar

Roads

The area has a good network of roads. Main roads, called 'Nacional' (marked N), connect the main towns and there are many provincial roads, joining the smaller towns, which are generally good. The smaller country roads are of varying quality but the majority are good for most vehicles. Some need re-surfacing and should be tackled with care especially after heavy rain in the winter. Some mountain roads are damaged by rock falls and landslides after prolonged bad weather and may be closed without prior warning.

The N340 between Málaga and Algeciras, which services the Costa del Sol, is heavily congested during the summer months and drivers on this road should exercise more than usual caution.

Unless otherwise specified, the speed limits in Spain are, 120 kph on motorways, 100 kph on main roads and 40 kph in built-up areas. The speed limit in Gibraltar, much of which is urban, is 40 kph. Seat belts are compulsory in Spain but they are not compulsory in Gibraltar.

Cars

The best way to travel is by car as it offers the bird-watcher flexibility and speed. This is particularly useful when trying to follow migration as it may be necessary to reschedule plans for the day quickly in response to changing weather.

National driving licences can be used in Gibraltar. In Spain it is advisable to carry an international driving licence even though it is not legally required for all nationalities. Citizens of the UK must be in possession of an international driving licence. Although green cards are not technically necessary for EC nationals it is advisable to be in possession of one when driving in Spain. Green cards can be purchased at border posts. A bail bond is recommended.

Vehicles are driven on the right-hand side of the road in Spain and Gibraltar.

Car Hire

All the established car hire firms operate in the area and those wanting to hire a car are advised to make a reservation in advance through their travel agent or direct with a car hire company. Cars can be collected at the airport. It is advisable to pay a little extra and have the vehicle insured for full comprehensive cover.

Taxis

Taxis are available in Gibraltar and most large Spanish towns. At the time of writing, there was no cross border taxi traffic between Gibraltar and Spain. Taxis can be hired at hotels and are especially useful for persons without a car who want to be taken and collected from a migration watching site. Be sure to agree a price before setting off on any lengthy journey.

Buses

Buses in Gibraltar service most of the peninsula and there are regular buses to Europa Point from the city centre and the airport. All Spanish towns are connected by a regular bus service. For some of the most frequently-used routes, such as Costa del Sol, it is advisable to book in advance at the local bus station.

Trains

The rail network in this area is not good and most of the sites cannot be reached by train. The Algeciras to Madrid line passes impressive mountain scenery between Algeciras and Ronda. The journey is worthwhile for the views and is one way to reach Ronda from the Strait area. Parts of the Costa del Sol are serviced by local trains. There are no trains in Gibraltar.

Hitch-hiking

Hitch-hiking prospects are good, especially along main roads. However, there is usually little advantage to be gained in hitch-hiking as buses follow the same roads and are cheap.

STAYING IN SOUTHERN SPAIN AND GIBRALTAR

Accommodation

Accommodation in Gibraltar is limited to hotels as there are no guest houses or camping facilities. There are eight hotels in Gibraltar. Prices (winter 1992/93) for a double room varied between £38 and £105 per night.

In Spain hotel accommodation varies from one star to five star. Most small towns have one star guest houses which are of varying, generally poor, quality. Prices vary substantially with location. In 1992, double rooms varied between 3,400 and 30,000 pesetas.

There are many camping and caravan sites in Spain especially along coastal areas. Although some are seasonal, many are open throughout the year. It is not advisable to camp away from official camping sites.

Places to eat

There is a good range of restaurants, coffee shops and fast food establishments in Gibraltar which cater for most tastes and pockets. Food ranges from fish and chips to exotic dishes. Most restaurants open at 12.00 for lunch and at 19.00 for dinner. There are many grocery shops and supermarkets in Gibraltar for those preparing their own meals and picnics.

In Spain there is a good range of restaurants in tourist-frequented areas along the coast but most mountain towns, especially the smaller ones, only have the traditional 'venta' which serves basic meals, and one or two small grocery shops. Spaniards eat late and allowances should be made for this in planning meals. It is not customary, for example, for lunch to be served before 13.00, and 14.00 to 15.00 is more usual. Dinner is usually eaten after 21.00.

Banks

There are many banks in Gibraltar including most well known British ones. Most are open from Monday to Friday between 09.00 and 15.30. Many banks re-open between 16.30 and 18.00 on Fridays. Cashpoint facilities are available.

In Spain most banks are open from 09.00 to 14.00 weekdays and 09.00 to 13.00 on Saturdays.

The main international credit cards and travellers' cheques are widely accepted in Gibraltar and Spain.

Safety and Theft

It is generally quite safe to walk around Gibraltar at any time, even at night. It is advisable not to walk alone, especially at night, in Spanish cities since there is a high incidence of crime. Avoid carrying valuables in view because of pick-pockets and snatch-and-grab specialists on motorbikes. When driving in some of the larger cities, keep valuables hidden in the boot as thefts can occur even in broad daylight while stopped at traffic lights. Sevilla is notorious for this.

Generally, there are no problems travelling or wandering in the countryside. Take care when bird-watching close to urban areas and do not leave anything exposed in the car while not in attendance. Always carry your passport and money with you.

Fire In the summer months (especially July-September) there is a high risk of forest fires in southern Spain. When travelling in the mountains make sure you are aware of your location and of all access and exit routes in case of an emergency. Behave responsibly in the woods. Do not light fires, drop cigarettes or discard glass bottles. Some tracks may be closed temporarily when there is a high risk of fire.

Bulls Some areas in southern Spain are fenced and may contain interesting habitats. If there are no signs prohibiting entry proceed with caution as there may be bulls present. Even the Retinto cows which are bred in the area can charge, especially when they have calves. These cows have the habit of grazing within wooded areas and can be quite invisible except at close quarters.

Coto Privado de Caza Many areas are marked with diagonal black-and-white square signs. These mean that the area is a private hunting estate and hunting usually only takes place at specific times of year. Rules for entry depend very much on the landowner but birdwatchers are usually tolerated. If a guarda or warden appears, do not avoid him but explain what you
are doing.

Penduline Tit

CLIMATE AND CLOTHING

Climate

The region has extremes of climate. The summer months are characteristically hot with summer temperatures inland exceeding 40°C. The coast is usually cooler with temperatures in the upper twenties and lower thirties. Winter temperatures are mild along coastal areas. Gibraltar has not recorded sub-zero temperatures during the last 40 years. The mountains can be very cold, especially the Sierra Nevada which has permanent snow throughout winter and spring. Most rain falls between November and April but the wet season is highly variable and can commence in September and finish in June. Some winters are fairly dry but others can be very wet with rain falling in heavy showers. Grazalema has the highest rainfall in Spain. The summer months are dry. A feature of the Strait area in particular is the prevailing high winds from the east or the west. Easterlies predominate in summer and westerlies in spring and winter. With easterlies, cloud usually forms over mountains, the best known being the 'Levanter' which forms over the Rock of Gibraltar.

Clothing

The weather patterns are variable and clothing depends very much on time and place. Even in summer the mountain tops can be cold, and the constant sea-breeze on the coast can feel chilly, especially when sea-watching for long periods. A sweater should be included even for a summer visit. Those planning to trek in summer should contemplate long trousers and boots despite the heat in order to avoid insect bites and abrasions from the many thorny plants which grow in the area. Always be prepared and take warm and waterproof clothing between September and May as dry spells are not guaranteed. Wellington boots are recommended for visits to lagoons and marshes.

HEALTH AND MEDICAL FACILITIES

The chances of catching a serious illness are remote. A few simple precautions should minimise the chance of minor illnesses spoiling a holiday. Contaminated food and water can cause problems, especially in summer when flies are abundant. Care should be taken not to eat food which has been left in the open for a long period. Mayonnaise should be avoided as it is prone to contamination. It is worth paying a little more for better food than buying cheaper food and risking illness. Tapwater should not be drunk in Spain but is safe to drink in Gibraltar.

If affected by diarrhoea, drink plenty of water with added salts which are available in sachets from chemists (available in the UK under the commercial name Dioralyte) and avoid greens and dairy products. If the symptoms persist for several days consult a doctor.

Visitors to the area should avoid long and continuous exposure to the sun, particularly in the summer. Use appropriate sun protection creams and wear a hat at all times. Never sleep in the sun and take care when sitting in one position for a long time (when watching raptor migration or sea watching) as the strong sun can burn very rapidly. Sunglasses are useful, especially when staring at the sky for long periods.

It is difficult for birdwatchers to avoid mosquito and other insect bites, especially when near marshy areas. Malaria does not occur in Spain and Gibraltar. Chemists sell a good range of insect repellants.

Many British pharmaceutical products are available in Gibraltar but it is always advisable to bring a selection of medications to cover likely eventualities.

Gibraltar and the larger Spanish cities have hospitals. There is also a Government Health Centre in Gibraltar. EC Nationals staying in Gibraltar and Spain may receive free hospital treatment on production of Form E111. United Kingdom citizens staying in Gibraltar can obtain free treatment within 30 days of leaving the UK.

Usually medical treatment is available if taken ill during the visit. Only immediately necessary treatment to enable the patient to recover is given.

MAPS

Road maps of Andalucía, published by Firestone and Michelin, are readily available and are useful for the traveller wishing to have all the sites on a single map. These maps are the ones most frequently used by visiting birdwatchers.

The Spanish Topographic Survey of the Servicio Geografico del Ejercito produce large scale maps (1:200,000, 1:100,000 and 1:50,000) with greater detail, including relief and other physical features. The 1:50,000 sheets are particularly useful for those intending to birdwatch away from the main roads.

Other detailed maps are available for some specific areas: a 1:50,000 map of the Sierra Nevada (Federacion Española de Montañismo); a 1:100,000 map of the Sierra de Cazorla (Editorial Everest); a 1:100,000 map of the Doñana National Park (ICONA); a 1:100,000 map of the Strait of Gibraltar (Instituto Geografico Nacional & Instituto Hidrografico de la Marina).

A selection of these maps is available in some Spanish bookshops and in the Geographic Institute offices in most Spanish cities. However, it is better to obtain them in advance both for planning a visit and to avoid delays once there. All the maps mentioned are available from Edward Stanford Ltd., 12-14 Long Acre, London WC2E 9LP [Tel: (071) 836 1321; Fax: (071) 836 0189].

Crested Coot

WHEN TO GO

Birds move through the area of the Strait throughout the year, making it difficult to divide the year by seasons. The bulk of migration takes place from March to May and from August to October but not all species conform to this pattern. White Storks, for example, pass south across the Strait in large numbers during late July and early August, and return north from the end of October. Any time of year can be rewarding for the visitor.

All year

Interesting species which can be found throughout the year include: Balearic Shearwater, Spoonbill, Greater Flamingo, Marbled and White-headed Ducks, Red Kite, Griffon and Black Vultures, Goshawk, Spanish Imperial, Golden and Bonelli's Eagles, Barbary Partridge, Purple Gallinule, Crested Coot, Little and Great Bustards, Avocet, Stone-curlew, Kentish Plover, Black-bellied and Pin-tailed Sandgrouse, Eagle Owl, Dupont's and Thekla Larks, Crag Martin, Alpine Accentor, Black Redstart, Black Wheatear, Blue Rock Thrush, Cetti's, Dartford and Sardinian Warblers, Crested Tit, Short-toed Treecreeper, Great Grey Shrike, Azure-winged Magpie, Chough, Spotless Starling, Spanish and Rock Sparrows, Serin and Hawfinch.

Apr-Sep

Those wanting to observe the greatest variety of species should visit during mid-April to mid-June. At this time birds are still migrating north and most of the summer visitors have arrived to breed.

Among the interesting summer visitors are: Little Bittern, Night, Squacco and Purple Herons, Black and White Storks, Black Kite, Egyptian Vulture, Short-toed Eagle, Montagu's Harrier, Booted Eagle, Lesser Kestrel, Black-winged Stilt, Collared Pratincole, Slender-billed Gull, Gull-billed and Whiskered Terns, Great Spotted Cuckoo, Scops Owl, Red-necked Nightjar, Pallid, Alpine and White-rumped Swifts, Bee-eater, Roller, Hoopoe, Short-toed and Lesser Short-toed Larks, Red-rumped Swallow, Tawny Pipit, Rufous Bush Robin, Black-eared Wheatear, Rock Thrush, Savi's, Great Reed, Olivaceous, Melodious, Spectacled, Subalpine, Orphean and Bonelli's Warblers, Woodchat Shrike, Trumpeter Finch and Ortolan Bunting.

Nov-Feb

The winter period, which may be considered to start in November and end in February, is also an interesting time as winter visitors from the north arrive and the resident species are still around. It is a good time to observe waterfowl, waders and some passerines in large numbers and some raptors such as Red Kite and Hen Harrier are most abundant at this time. The winter is a suitable time for a short visit and the weather is usually mild except in the mountains.

Interesting wintering species are: Greylag Goose, Red-crested Pochard, Osprey, Crane, Black-tailed Godwit, Mediterranean Gull, Alpine Accentor and Penduline Tit.

Migration watching

For spectacular migratory movements visits should be made during the main passage periods, although the species to be seen depends very much on the month.

Feb-Mar

February to March is good for Cory's Shearwater, Black Kite and Short-toed Eagle.

Mar-Apr

During late March to April there is a good variety of passage raptors including Black Kite, Egyptian Vulture, Short-toed Eagle, Marsh and Montagu's Harriers, Buzzard, Booted Eagle and Osprey.

Apr-May

April and May are excellent for migrating passerines and large falls occur in the Strait area after bad weather.

Jul-Aug

July and August are probably the best sea-watching months. There are large numbers of Balearic and Cory's Shearwaters offshore and there is a large westward movement of thousands of Audouin's Gulls which pass close to the shore. Other seabirds are likely at this time. July to August is also the peak of the southward migration of the White Stork, Black Kite and Swift.

Aug-Oct

The last week of August and the first ten days of September are best for large numbers of raptors on migration, especially Honey Buzzard, Egyptian Vulture and Montagu's Harrier. Black Storks and Short-toed and Booted Eagles follow at the end of September and in early October. The main passerine activity in autumn is in late September to mid-November when large falls occur after bad weather. Finch migration during October and early November is spectacular.

Honey Buzzard

INTRODUCTION TO SITE INFORMATION

The site information is arranged to maximise bird-watching time and reduce travelling time. The large size of Andalucía makes it difficult for birdwatchers to comprehensively explore the region during a short visit. Most species typical of the region can be found in Gibraltar, Cádiz and western Málaga. The area includes the Strait of Gibraltar which lies on a major migration route. Within this area the landscape is varied: in the west there are lowlands with marshes, sandy beaches and lagoons; in the north and east are impressive mountains and extensive woods. The coastal hills of the Strait are covered in woodland and scrub and the Strait itself adds seabirds to the list of potential species.

Sites which are close to each other are grouped together and in choosing sites, care has been taken to avoid excessive duplication. An excellent variety of migrants can be seen within 25 km of Gibraltar. The migration sites within the Strait area have been selected to provide a range of these. Raptor migration is heavily influenced by wind direction and force so the best lookouts can vary even within a day. During strong westerlies the Rock of Gibraltar is best but during easterlies the birds fly further west over Tarifa. Europa Point and Punta Carnero are recommended for seabird migration. The scrub at Gibraltar and Punta Carnero, the Jara Valley and the pine woods around Cape Trafalgar is excellent for migrant passerines. The Guadiaro and Palmones estuaries also attract passerines and a good range of waders and waterbirds.

Two important sites, some distance from the main nucleus of the other sites, are included - the Coto Doñana and Cabo de Gata. There are brief notes on other sites spread across Andalucía for those who venture further or are travelling to the area from the north by car.

The following species are not mentioned in the bird lists given for each site. They are all common and widely distributed in the appropriate habitat. Most are resident, few are migrant and their status is given in the full bird list (p74).

Little and Great Crested Grebes, Cormorant, Cattle and Little Egrets, Grey Heron, Wigeon, Gadwall, Teal, Mallard, Shoveler, Pochard, Sparrowhawk, Buzzard, Kestrel, Pheasant, Moorhen, Coot, Oystercatcher, Ringed and Grey Plovers, Lapwing, Dunlin, Snipe, Curlew, Redshank, Common Sandpiper, Turnstone, Black-headed, Lesser Black-backed and Yellow-legged Herring Gulls, Sandwich, Common and Little Terns, Woodpigeon, Turtle Dove, Cuckoo, Barn and Little Owls, Swift, Kingfisher, Great Spotted Woodpecker, Skylark, Sand Martin, Swallow, House Martin, Meadow Pipit, Yellow, Grey and White Wagtails, Wren, Robin, Wheatear, Blackbird, Song and Mistle Thrushes, Sedge and Reed Warblers, Whitethroat, Garden Warbler, Blackcap, Chiffchaff, Willow Warbler, Spotted Flycatcher, Blue and Great Tits, Nuthatch, Jay, Magpie, Jackdaw, Starling, House Sparrow, Chaffinch, Greenfinch, Goldfinch, Linnet and Reed and Corn Buntings.

The Rock of Gibraltar

The Rock of Gibraltar is a limestone peninsula, 5km by 2km, at the southernmost tip of Iberia. The town, first colonised by the Moors in 711 AD, is situated at the western base of the spectacular Rock which towers to an altitude of 426m at O'Hara's Battery. The western slopes of the Rock are covered in Mediterranean scrub (maquis) dominated by wild olives and other typical shrubs. Towards the southernmost tip, Europa Point, is a series of terraces covered in low scrub. The most important of these terraces is Windmill Hill. The northern and eastern sides of Gibraltar are sheer cliff faces with scree slopes at the bases. A characteristic feature of Gibraltar is the number of caves which are found scattered throughout the peninsula. Some, such as the isolated sea caves at Governor's Beach, are important roosts for large numbers of Crag Martins. The whole of the Rock is considered an Important Bird Area in Europe by the International Council for Bird Preservation.

The main areas for bird-watching on Gibraltar are the Upper Rock, Windmill Hill, Europa Point and the small cemetery at the northern end of the peninsula. Interesting migrants can be found in any location within Gibraltar (e.g. the gardens around the town itself), especially after bad weather has produced a fall. Spectacular raptor movements may be observed from any point on Gibraltar but the recommended points are the best places to obtain good views of large numbers of birds.

Key

Lighthouse	
Roads	
Tracks	
Cliff	

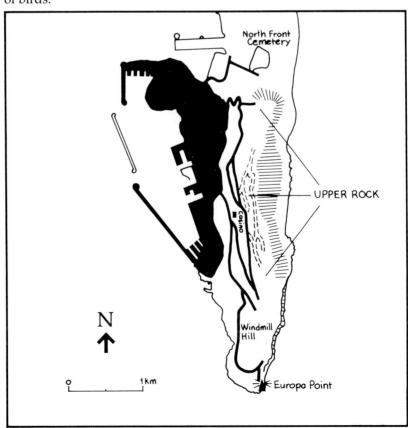

Gibraltar makes an excellent base for a bird-watching trip, having an airport and a good range of hotels. Many of the sites included in this guide can be visited on day trips from Gibraltar and birdwatchers planning a visit to the area, especially if interested in observing migration, should consider using Gibraltar as a base and hiring a car. Information on Gibraltar in general and accommodation in particular can be obtained from the Gibraltar Information Office at Market Place and 18-20 Bomb House Lane, Gibraltar (Tel. Gibraltar 74982).

Upper Rock

The Mediterranean maquis which covers the west side of the higher portions of the Rock has a rich flora containing some species which are either endemic to Gibraltar (e.g. Cerastium gibraltaricum) or unique in Europe (e.g. Iberis gibraltarica). The scrub attracts migrant passerines and near-passerines, often in large numbers. Migrating raptors and small migrants (hirundines, swifts, bee-eaters, etc.) fly over the area in large numbers during the day.

Strategy   The Upper Rock is best visited during passage periods (February-early June and late July-November) although there is also an interesting wintering bird community. For diurnal passage it is best to watch from one of the recommended vantage points in westerly winds only, when birds are drifted overhead. Passage of small migrants takes place early in the morning and late in the evening. Raptors pass throughout the day and are lower in the morning and evening. Most of the heavier species (e.g. Short-toed Eagle, Griffon Vulture) pass during the warmest part of the day. Early morning is the best period for walking along the Upper Rock in search of grounded migrants. For these, overcast conditions produced by easterly winds are best.

There are several points of access to this area but the recommended route, by car or on foot, is from Europa Road. Take a left turning just past the Casino and follow this road which heads south. This area is a nature reserve and there is an entry charge of £3 per person and £1.50 per car. Contact the local information offices in advance for information on special concessions for bona fide birdwatchers. Once in this area there are many paths and roads which can be followed. The area can be covered in a morning's walk although a whole day can easily be spent on this site during the migratory periods. At such times the site can be visited on consecutive days as the turnover of migrants is high. Three vantage points are recommended for observation of visible migration: Princess Caroline's Battery and Signal Station for the southbound migration and Jews' Gate for the northbound migration. From these points in suitable weather it is possible to get very close to migrating raptors and other diurnal migrants. Much of the Upper Rock is attractive to passage migrants which rest and feed here although the dense cover of vegetation makes observation difficult at times. The roads and paths along the higher areas of the Upper Rock (above St Michael's Cave) are usually the most productive although the lower slopes around Jews' Gate are also favoured by migrants. A walk along

Mediterranean Steps is recommended for its scenic beauty and also provides close views of Peregrine Falcons, and Barbary Partridges are usually found here. This is a good site for Alpine Accentor in winter.

Birds   Black and White Storks, Honey Buzzard, Black and Red Kites, Egyptian and Griffon Vultures, Short-toed Eagle, Marsh, Hen and Montagu's Harriers, Booted Eagle, Osprey, Lesser Kestrel, Hobby, Peregrine (R), Barbary Partridge (R), Great Spotted Cuckoo, Scops Owl, Nightjar, Red-necked Nightjar, Alpine Swift, Bee-eater, Hoopoe, Wryneck, Short-toed Lark, Crag Martin, Red-rumped Swallow, Tawny Pipit, Alpine Accentor (W), Rufous Bush Robin, Black Redstart, Redstart, Whinchat, Black-eared Wheatear, Rock Thrush, Blue Rock Thrush (R), Ring Ouzel, Olivaceous, Melodious, Dartford, Spectacled, Subalpine, Orphean and Bonelli's Warblers, Firecrest, Pied Flycatcher, Short-toed Treecreeper, Golden Oriole, Great Grey and Woodchat Shrikes, Serin, Siskin, Rock and Ortolan Buntings.

Barbary Partridge

Other Wildlife   Barbary Macaque, Schreiber's Bat, Large Mouse-eared Bat, Large Psammodromus, Iberian Wall Lizard, Moorish Gecko, Horshoe Whip Snake and Montpellier Snake.

Windmill Hill

Windmill Hill is a raised platform at the southern end of the Gibraltar peninsula. It is the stronghold of the Barbary Partridge on Gibraltar and a prime site for migrants. This 13 hectare site is surrounded on its eastern, southern and western perimeters by lime-stone cliffs. The vegetation is essentially coastal garrigue with areas of steppe and patches of taller scrub. There are a number of military buildings and other installations within the site, which is used as a training ground.

Strategy To reach Windmill Hill, follow Europa Road towards Europa Point. There is a left turning just before reaching the Naval Hospital which leads to a military post. The area is military property and access is restricted. It may be possible, however, to obtain permission to visit the site but it is advisable to apply well in advance. Permission should be sought from The Adjutant, The Gibraltar Regiment, Lathbury Barracks, Gibraltar. The permit must be presented at the military post to obtain access.

A small patch of maquis at the western end of Windmill Hill, below the perimeter cliff, known as Jacob's Ladder is open to the public. Access is from Europa Road near St Bernard's Chapel. A footpath leads through the scrub to the base of the cliff.

Barbary Partridge can be easily found at Windmill Hill throughout the year with large coveys forming between July and January. Many (over 40 have been recorded together) roost on the western cliffs and can be seen from Jacob's Ladder at sunset, especially at the end of the breeding season (August-October). Other resident species include Blue Rock Thrush and Sardinian Warbler whilst Peregrine Falcons which breed nearby often hunt over the site.

Otherwise, Windmill Hill is essentially a place to look for migrant passerines between February and early June and between late July and November. The greatest concentrations occur after periods of easterly winds which produce overcast conditions, sometimes with rain. During westerly winds it is possible to find some grounded migrants and, in spring, migrating raptors fly very low over the site.

Birds Windmill Hill is probably the best site in the whole region for variety and concentration of small migrants. It attracts more open ground species than the nearby Upper Rock. The species arriving at the site vary greatly according to time of year and weather conditions and many accidentals have been recorded here in recent years. These have included Desert Wheatear, Tristram's Warbler, Trumpeter Finch, Pine Bunting and Bobolink.

Regular migrants in spring and autumn include Black and White Storks, Honey Buzzard, Black Kite, Egyptian and Griffon Vultures, Short-toed Eagle, Marsh, Hen and Montagu's Harriers, Booted Eagle, Osprey, Lesser Kestrel, Hobby, Quail, Great Spotted Cuckoo, Nightjar, Red-necked Nightjar, Alpine Swift, Bee-eater, Hoopoe, Wryneck, Short-toed Lark, Woodlark, Crag Martin, Red-rumped Swallow, Tawny Pipit, Rufous Bush Robin, Black Redstart, Redstart, Whinchat, Black-eared Wheatear, Rock Thrush, Ring Ouzel, Redwing, Olivaceous, Melodious, Dartford, Spectacled, Subalpine, Orphean, Bonelli's and Wood Warblers, Pied Flycatcher, Golden Oriole, Great Grey and Woodchat Shrikes, Serin, Siskin and Ortolan Bunting.

Other Wildlife Ocellated Lizard (scarce), Iberian Wall Lizard, Moorish Gecko, Large Psammodromus, Horshoe Whip Snake, Southern Smooth Snake.

Europa Point

The southernmost tip of Gibraltar consists of a built-up area of flat ground with a lighthouse and a promenade on the edge of 20 metre limestone cliffs which reach the shores of the Strait of Gibraltar. At the south-western corner, below the cliff line, a flat area of foreshore is covered in semi-natural low coastal scrub which includes the endemic sea lavender. This is primarily a sea-watching site from which the movements of seabirds between the Atlantic Ocean and the Mediterranean Sea can be easily observed. The point, however, attracts many other passage migrants including waterbirds, raptors and passerines so that it is often possible to see many different types of birds while sea-watching.

Strategy   Europa Point is best visited in late afternoon and evening when there is most seabird activity and light conditions are best for sea-watching. Onshore winds (south-west or south-east) are best for bringing seabirds close to the shore. Passage periods (February-June and July-October) and winter are the best times. May is the poorest month for sea-watching.

From the city centre take Europa Road in a southerly direction until the Europa Point lighthouse becomes visible. Follow the road towards the lighthouse. There is a parking area beside a small tea-shop. From here walk towards a small vegetated mound, west of the lighthouse from where it is possible to view the Strait. In strong westerly winds it is possible to shelter in the lee of a small building on the edge of the cliff. The foreshore can be reached by a wooden staircase to the north-west of this mound.

The small mound and the foreshore attract small migrants which usually disperse towards the Upper Rock after late morning. A walk around these vegetated areas is worth trying. Apart from the typical migrants (see Windmill Hill) the area has recently attracted accidentals such as Arctic, Yellow-browed and Dusky Warblers and Trumpeter Finch. In spring especially, it is possible to obtain excellent views of migrating raptors flying low over the sea as they arrive from Africa.

Birds   Most typical North Atlantic and Mediterranean seabirds can be seen from Europa Point. Among the most noteworthy movements are the westward passages of Audouin's Gulls in July and August (several thousand) and Puffins, Razorbills and Gannets, in March and April. Cory's Shearwaters are regular, sometimes in large numbers, during June to September with spectacular migratory movements in October/November (west) and February/March (east). Balearic Shearwaters are present most of the year but peak movements occur in June/July (west) and December/February (east). In winter, large con-centrations of seabirds are found off this point after bad weather, including Little and Mediterranean Gulls and Kittiwake.

Other interesting seabirds which are regular off Europa Point are Shag (resident Mediterranean sub-species), Arctic and Great Skuas, and Black Terns.

Regular species in small numbers during passage periods include Sooty Shearwater (September), Pomarine Skua (May, August-September), Gull-billed Tern (April, July-August), Royal Tern (August-September) and Lesser Crested Tern (October). Recent accidentals have included Little Shearwater, Ring-billed, Laughing, Iceland and Sabine's Gulls and Little Auk.

Other Wildlife   Leathery Turtle (occasional), Loggerhead Turtle (occasional), Common Dolphin (abundant), Long-finned Pilot Whale, Killer Whale (occasional).

North Front Cemetery

North Front Cemetery is situated at the southern corner of the sandy isthmus which separates the northern end of the Rock of Gibraltar from Spain. Take Devil's Tower Road eastwards and then a left turning which leads to Cemetery Road. The cemetery has areas with hedges and some tall trees (Eucalyptus), particularly at the north-eastern corner. It attracts passerine and near-passerine migrants in spring and autumn.

Strategy   The area attracts many migrants during the migration periods. It is at its best early in the morning when there is little disturbance from the public and newly arrived migrants can be found actively feeding. Lesser Kestrels and Peregrines which nest on the tall cliffs of the north face of the Rock regularly fly over the site.

Birds   Regular migrants at this site, sometimes in large numbers during falls, include Quail, Great Spotted Cuckoo, Scops Owl, Nightjar, Red-necked Nightjar, Bee-eater, Hoopoe, Wryneck, Tawny Pipit, Rufous Bush Robin, Black Redstart, Redstart, Whinchat, Black-eared Wheatear, Olivaceous, Melodious, Dartford, Spectacled, Subalpine, Orphean and Bonelli's Warblers, Pied Flycatcher, Golden Oriole, Woodchat Shrike, Serin, Siskin and Ortolan Bunting.

Other Wildlife   Ocellated Lizard, Iberian Wall Lizard, Large Psammodromus, Praying Mantis.

Algeciras Sierras

Alcornocales Natural Park (south)

Five sierras form the backbone of the Spanish side of the Strait of Gibraltar - Cabrito, Bujeo, Algarrobo, Ojén and Luna. They form a mass of sandstone which reaches the shores of the Strait between Punta del Carnero and Tarifa. To the north of these sierras other sierras meet with the sandstone mountains of the Aljibe (1092m). These mountains are unique in providing a large area of relatively undisturbed Cork Oak woodland with patches of Maritime Pine woodland and open pastures. The sandstone sierras are better vegetated than the exposed limestone of the Serranía de Ronda (p.51) and large cliffs are not found. Instead, small and scattered rocky outcrops occur providing nesting and resting places for raptors.

The western sides of the mountains face the Atlantic and gather rain clouds during the winter months. Rainfall is high and the mountains and valleys are bisected with numerous fast-flowing streams. From here the rivers Miel and Picaro flow to the Bay of Gibraltar.

The dominant trees are Cork and Algerian Oaks, with Alder close to the mountain streams. Around the Sierra del Cabrito there are copses of Maritime Pine. A number of interesting plants grow in these mountains, including wild rhododendrons and foxgloves as well as the insect-eating *Drosophilum lusitanicum*. The higher, scrub covered areas are dominated by heathers, rock roses and brooms.

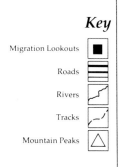

Key

Migration Lookouts

Roads

Rivers

Tracks

Mountain Peaks

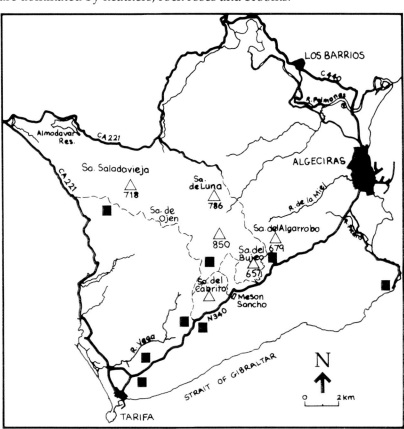

All of the sierras are of similar ornithological importance, being used by migrating birds in the spring and autumn, and by a rich community of breeding birds. The secluded valleys are used by large numbers of migrating raptors as roosts. The rocky shores of the Strait attract small numbers of waders and sea-watching is possible from vantage points along the coast.

The highest sierras are Ojén (810m) and Luna (786m) which form the backbone of the area. Cabrito (536m) faces south-west, Bujeo (657m) south, and Algarrobo (679m) south-east.

Accommodation

The area is within easy reach of Gibraltar so that those based on the Rock can comfortably drive to the sierras in less than an hour. Closer accommodation is available in Algeciras which has a good range of hotels from 1-star to the 4-star Reina Cristina. Prices for a double room range from 2,000 to 14,500 pesetas per night. The Meson de Sancho (2-star) is well placed in wooded surroundings by the N 340 at km94. Another option is to stay near Tarifa.

Strategy

The sierras are worth a visit at any time of year. February-June is good for breeding birds and spring passage, late July-early November for autumn passage and November-February for wintering birds.

Most of the breeding species of Cork Oak woodland and upland scrub and pastures can be found here. The close proximity to the Strait makes it an excellent landfall for small migrants while migrating raptors fly through, often very low, having just completed the sea crossing.

In late summer and autumn many raptors roost in the woodlands before departing towards the Strait and Morocco. Together with the Rock of Gibraltar, the Algeciras Sierras form the main front for raptors crossing the Strait at both seasons. While Gibraltar is best in fresh westerlies, Algarrobo is ideally located for passage in light westerlies, whereas Ojén and Cabrito are best in easterlies, the precise path being determined by the wind strength. In the strongest easterlies in late summer and autumn, most raptors do not cross the Strait but wander over the sierras. In continuing easterlies, large flocks can often form.

The sierras can be approached from a variety of directions and are criss-crossed with a network of firm tracks and paths. To approach them from the north, take the CA 221 which branches east from the CA 440, 6km north of Los Barrios. The CA 221 is a scenic 25km country road which is worth a morning on its own. The woods along this road are good for birds and there are some mature stands of Cork Oak and Algerian Oak along this stretch. The small Almodovar Reservoir at the western end of the road marks the boundary of the Algeciras hills and, though not usually good for birds, it can attract some migrating water-birds and waders.

Travelling west along the CA 221 there is a left turn 9km from the junction with the CA 440. This road eventually becomes a track which crosses the sierra and joins up with the N 340 coast road. The track is firm in summer and most of it can be driven on, although a four-wheel drive is recommended for the worst stretches. In winter the surface deteriorates after rain and cars should not be used after the first transmitter has been reached. This track runs south and eventually links up with a road which follows the eastern part of the sierras between Bujeo and Algarrobo, meeting the N 340 between km95 and km96. This junction can be used as an access point from the N 340.

Southern parts of the sierras can be approached from the N 340 although there are several gates across the tracks which can be opened but must be closed afterwards. One access point is through a small residential complex (El Cuarton) between km93 and km94 and another is just west of Puerto del Cabrito at km91. All these access points are on the north side of the N 340 and they meet in the sierras and join with the northern track. It is possible to follow the track north-west which links up with the CA 221 5km north of Santuario de Nuestra Señora de la Luz.

It is possible to stop at any point and walk in search of birds. A map and compass is recommended when walking any distance from the track as low cloud can descend quickly in the mountains. A walk to the highest point, Tajo de la Corza, is scenic and the view from the summit is spectacular.

Care should be taken in trying to gain access to the coastal side of these sierras since much of it is military ground. Signs should always be respected. The Punta Carnero to Punta Secreta area is accessible and worth a visit. Take the CA 223 from km103 on the N 340 (there is a sign indicating Getares). This scenic country road follows the rocky south-western coast of the Bay of Gibraltar past the village of Getares, reaching just beyong the Punta Carnero lighthouse (8km from the junction with the main road) to a rocky point known as Punta Secreta. Sadly, part of this area is being developed but it is still worth a visit. There is a good cover of coastal scrub dominated by brooms and the rocky shore attracts passage waders and seabirds. Flocks of Audouin's Gulls can be seen resting on the rocks in summer and Purple Sandpipers are regular in winter (a very southerly site for this species). Seabird movements can be followed from here and migrating raptors pass overhead in light westerlies, often very low in the spring. The reed-covered banks of the Rio Picaro alongside Getares are worth a look as they can attract passage waterbirds.

Birds

Interesting residents of the sierras include ; Griffon Vulture, Goshawk, Bonelli's Eagle, Peregrine, Eagle and Tawny Owls, Green Woodpecker, Thekla Lark, Woodlark, Dipper, Blue Rock Thrush, Dartford Warbler, Firecrest, Long-tailed Tit, Crested Tit, Short-toed Treecreeper, Raven, Spotless Starling, Serin, Hawfinch and Cirl and Rock Buntings.

Summer visitors which breed include: Egyptian Vulture, Short-toed and Booted Eagles, Lesser Kestrel (in Los Barrios), Scops Owl, White-rumped Swift, Bee-eater, Red-rumped Swallow, Tawny Pipit, Black-eared Wheatear, Olivaceous, Melodious and Bonelli's Warblers, Golden Oriole and Woodchat Shrike.

A variety of migrants can be seen in the area. Some use the woodland and other vegetated areas and many converge over the Strait. Seabirds and waders can be seen along the shore. Highlights among this range of species include: Cory's and Balearic Shearwaters, Honey Buzzard, Black Kite, Egyptian and Griffon Vultures, Short-toed Eagle, Marsh, Hen and Montagu's Harriers, Golden Eagle (occasional on passage and in winter), Booted Eagle, Osprey, Lesser Kestrel, Hobby, Purple Sandpiper (regular in winter), Woodcock (winter), Whimbrel, Green Sandpiper, Great Skua, Mediterranean and Audouin's Gulls, Black Tern, Razorbill, Puffin, Bee-eater, Hoopoe, Redstart, Whinchat, Black-eared Wheatear, Ring Ousel, Subalpine Warbler and Pied Flycatcher.

White-rumped Swift

Other Wildlife   Ocellated Lizard, Ladder Snake, European Pond/Stripe-necked Terrapin, Bosca's Newt, Genet, Red Fox, Weasel, Western Polecat, Egyptian Mongoose, Red Deer, Common Dolphin and Long-finned Pilot Whale.

Valleys of Guadiaro, Genal, Hozgarganta, Guadarranque and Palmones

The Guadiaro, Guadarranque and Palmones are the three largest rivers in the proximity of Gibraltar. The Genal and the Hozgarganta are tributaries of the Guadiaro. The Guadiaro opens into the Mediterranean Sea about 14km north-east of Gibraltar. The Palmones and Guadarranque open into the Bay of Gibraltar.

The three estuaries are ornithologically interesting, being strategically placed to attract birds moving along the Mediterranean coastline. They have been largely overlooked by visiting bird-watchers. The lower reaches of the Guadiaro, with shallow banks and gravel islands, contrast with the steep-sided valleys of the lower Hozgarganta near Jimena de la Frontera. The estuaries and the lower parts of the Guadiaro and Hozgarganta, together with the interesting lowland woods of Almoraima and Pinar del Rey, are highlighted in this section.

Key

Roads	
Rivers	
Mountain Peaks	

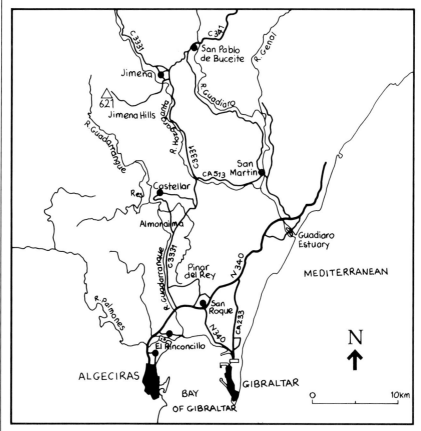

All the sites described are within a short drive from Gibraltar, the furthest being the Jimena hills which are 45km from Gibraltar. The other main towns adjacent to the sites are Algeciras, San Roque and Jimena de la Frontera.

Strategy   These sites are interesting throughout the year and also attract migrants in spring and autumn. The habitats (freshwater, marsh, reed beds, riverside woodland) are different from other habitats in the area and are likely to produce migrants which are less frequent elsewhere. They therefore complement Gibraltar, the Tarifa area and the Algeciras sierras. These sites also hold some interesting breeding and wintering species.

A day itinerary in spring could start with the Guadiaro Estuary, followed by the San Martin area which can eventually lead to Jimena and a visit to the hills around the Hozgarganta. If based in Gibraltar or Algeciras, then Almoraima, Pinar del Rey and Palmones can be done on the return journey.

Accommodation   Since all sites are close to Gibraltar, the visitor staying on the Rock would not require separate accommodation. There are hotels in Algeciras (see Algeciras sierras, page 21). Jimena and San Roque have a few 1-star and 2-star hotels/hostels. There are camping facilities on Km124 of the N 340 near San Roque .

Lower Guadiaro and estuary

The Guadiaro opens into the Mediterranean north-east of Gibraltar. It is one of the two largest estuaries on the coast between Gibraltar and Málaga, the other being the Guadalhorce, near Málaga. It is well placed to attract birds flying along the Mediterranean coast to and from the Strait of Gibraltar. The Guadiaro is wide near the sea but a sandy spit closes the estuary considerably where it reaches the Mediterranean. The approaches to the estuary are wooded, particularly with White Poplars, and backed by cultivated fields. Much of the marsh which once surrounded the estuary has been lost to the Sotogrande development, an up-market residential complex.

Strategy   The site is well worth a visit from Gibraltar which is about 15km away and could be included in an itinerary together with the Palmones and Guadarranque or on a half day visit on its own.

The estuary is best during the migratory periods and in winter and it is possible to visit it on consecutive days and see different species as the turnover of birds is high. Morning and evening are best since birds are most active and disturbance is least although any time of day can be good.

Despite the obvious signs of development all around the site, the Guadiaro Estuary is an important site for birds and well worth a visit. The estuary itself and its south-western bank still have reed beds and stands of Tamarisk which attract many interesting birds. The sandy beach is good for waders, especially near the estuary, and the sea area immediately offshore is worth a look for seabirds and wildfowl. At the north-eastern end of the site, the small lagoon near Torreguadiaro attracts waterbirds and other migrants. In winter, many of the surrounding fields are flooded and worth checking for waders.

A few kilometres upstream, the Genal, a tributary of the Guadiaro, joins it. The river is slow-flowing and there are many gravel banks and islands on which Little Ringed Plovers breed and other freshwater waders occur on passage. It is an area frequented by Ospreys, sometimes even in mid-summer. The orange and olive groves in the area attract migrants and Orphean Warbler and Rufous Bush Robin are breeding species.

The area around San Martin del Tesorillo is interesting during the breeding season. It has a good breeding population of Little Ringed Plovers and Bee-eaters are abundant. Raptors from the nearby mountains, including Bonelli's Eagle, hunt close to the river.

Take the turning towards the coast (east) marked Sotogrande at Km132 on the N 340. There is a security post at the entrance to the

Sotogrande development but access is not prohibited. Follow the road through the estate which leads to the estuary itself. Walk along the beach towards the estuary. Return to the car and drive northwards until you reach a bridge which crosses the river. From the bridge you can observe the estuary from a slight elevation as well as the vegetated area on the right bank which can be walked along a series of paths. From here continue north-east through the development, past the marina, and on the right is the lagoon at Torrguadiaro (an old Moorish tower). Park the car and walk around the lagoon. Return to the car and join the main road travelling back towards Gibraltar. In less than 1 km, look for a track on the left (this should be before the right turning to San Enrique de Guadiaro). In dry weather it is possible to take a car down the track although a 4-wheel drive car would be better. Follow the track (preferably on foot as birds can be seen on the way) which leads to the river and a good stretch of riverside woodland.

If instead, the right turning to San Enrique is followed, this country road will eventually lead to San Martin del Tesorillo (CA 513). Just before San Martin there is a country road on the right (in poor condition) which should be followed. It leads to the shore of the Guadiaro and the Genal in a number of places and a walk upstream is usually rewarding. Those wanting to follow the river further should proceed past San Martin. The road, which is in a bad state, follows the Guadiaro and joins the main road (C 341) close to San Pablo de Buceite, near Jimena de la Frontera.

Birds   Interesting species using the site on passage and/or in winter are: Black-necked Grebe, Squacco and Purple Herons, Greater Flamingo, Shelduck, Pintail, Pochard, Common Scoter, Osprey (wintering site), Black Kite, Booted Eagle, Marsh, Hen and Montagu's Harriers, Purple Gallinule, Water Rail, Little Ringed and Kentish Plovers, Little Stint, Curlew Sandpiper, Knot, Sanderling, Ruff, Spotted Redshank, Greenshank, Green and Wood Sandpipers, Black-tailed and Bar-tailed Godwits, Whimbrel, Jack Snipe, Black-winged Stilt, Avocet, Stone-curlew, Collared Pratincole, Mediterranean, Little and Audouin's Gulls, Caspian, Whiskered and Black Terns, Nightjar, Red-necked Nightjar, Short-eared Owl, Bee-eater, Hoopoe, Short-toed Lark, Crag Martin, Red-rumped Swallow, Tawny Pipit, Woodchat Shrike, Cetti's, Great Reed and Melodious Warblers, Pied Flycatcher, Whinchat, Black-eared Wheatear, Black Redstart, Redstart, Bluethroat (wintering site), Penduline Tit (wintering site), Reed Bunting and Serin.

Other Wildlife   Genet, Egyptian Mongoose, Stripe-necked Terrapin, Viperine Snake.

Jimena Hills

West of the town of Jimena de la Frontera, a range of hills faces the Strait of Gibraltar and the Atlantic. At the base of these hills, the Hozgarganta River flows past Jimena itself. The lower parts of the hills are devoted to goat and cattle pasture but higher up there are large tracts of Cork Oak woodland. The peaks are covered by a dense scrub dominated by heathers, rock roses and Strawberry Trees. There are areas of exposed sandstone near the tops and some interesting caves which are used by nesting Crag Martins. The hills reach a maximum altitude of 621m and dominate the surrounding lowlands. Being in a direct line north of the Strait, they are in the flyline of diurnal migrants. Migrating raptors can be seen following the ridges to gain lift from upcurrents, sometimes pausing to rest or hunt. The area is within the Alcornocales Natural Park.

Strategy   This site is best visited in spring (February-June) when a combination of breeding birds and passage migrants can be seen. Winter is a quiet period and the summer months should be avoided due to high temperatures, risk of forest fires and because the species present then can be readily observed at other sites.

From Jimena, take the country road which leaves the town in a south-westerly direction and crosses a small bridge over the River Hozgarganta. From here the road can be followed for 10km to a transmitter at the highest point. It is possible to stop in many places along the road and wander into the hills. A walk along the Hozgarganta itself from the bridge can be rewarding. The area around the peak is usually good for resident and migrating raptors as well as spectacular views of Gibraltar, the Strait, North Africa and the Atlantic.

Alternatively, the Hozgarganta can be reached by taking the C 3331 towards Ubrique. This road crosses interesting Cork Oak woodland and the river can be reached from numerous spots.

Birds   The following species are of interest in the recommended period: Night Heron, White Stork (several nests along C 3331 south from Jimena to Estación de San Roque), Honey Buzzard, Black Kite, Egyptian and Griffon Vultures, Short-toed Eagle, Marsh, Hen and Montagu's Harriers, Goshawk, Booted and Bonelli's Eagles, Osprey, Lesser Kestrel, Hobby, Peregrine, Green Sandpiper, Scops, Eagle and Tawny Owls, White-rumped Swift, Bee-eater, Hoopoe, Woodlark, Crag Martin, Red-rumped Swallow, Tawny Pipit, Black Redstart (early spring for wintering individuals), Black-eared Wheatear, Blue Rock Thrush, Ring Ouzel, Cetti's, Melodious, Dartford and Bonelli's Warblers, Firecrest, Crested Tit, Short-toed Treecreeper, Golden Oriole, Woodchat Shrike, Raven, Serin, Hawfinch and Cirl and Rock Buntings.

Other Wildlife   Red Deer, Genet, Fox, Western Polecat, Otter.

Almoraima

The Almoraima, traditionally known as the Cork Woods, is a large 16,000 hectare estate. It has escaped the development which has ruined many lowland areas and the traditonal industries have been the collection of the bark of the Cork Oak and hunting. Today the Almoraima is state run and it is possible to stay at the old convent and obtain permission to visit some of the remote parts of the estate. The Almoraima abounds with woodland birds at all times of year but the best time is undoubtedly spring when residents and summer visitors are most active and there is the chance of finding a few migrants.

Accommodation   Gibraltar and Algeciras are close to Almoraima. Alternatively, accommodation is available in the Casa Convento on the estate itself.

Strategy   Almoraima can be visited as part of an itinerary covering other near-by sites or it could be visited for a day on its own. The area is large and it is possible to stay at the Casa Convento from where permission can be sought to visit the remote and restricted parts of the estate. Any time of year is worthwhile but spring is recommended.

Take the C 3331 which runs between Estación de San Roque and Jimena de la Frontera. Driving from Jimena, there is a right turning after about 22km marked Castellar. Just before this there is a turning which leads to the convent which is visible on the right. This turning should only be taken by those wishing to stay at the convent. Otherwise, follow the road marked Castellar which runs through the Almoraima woods and, as the road climbs towards the old town of Castellar with its castle on the top of a hill, through rock rose dominated scrub. Castellar is 7km from the junction of the C 3331 and the road descends beyond Castellar to rejoin the main road. Instead of following the road to Castellar, it is possible to take a left turning after 5km which continues along woodland and leads to a lookout overlooking a large reservoir. There are many potential stopping points in these woods from which to watch and wander, although many areas are fenced.

Birds   Interesting breeding species in spring include Griffon Vulture, Short-toed Eagle, Goshawk, Booted Eagle, Lesser Kestrel (colony in Castellar itself), Scops Owl, Nightjar, White-rumped Swift, Bee-eater, Woodlark, Crag Martin, Red-rumped Swallow, Black-eared Wheatear, Blue Rock Thrush, Olivaceous, Melodious and Bonelli's Warblers, Firecrest, Long-tailed Tit, Crested Tit, Short-toed Treecreeper, Golden Oriole, Woodchat Shrike, Raven, Spotless Starling, Serin, Hawfinch and Cirl and Rock Buntings.

Other Wildlife   Red Deer, Egyptian Mongoose, Genet, Otter, Fox, Western Polecat, Weasel.

Pinar del Rey

The Pinar del Rey is a large Stone Pine wood situated between the rivers Guadarranque and Hozgarganta/Guadiaro on high ground. It is the largest pine wood in the area and parts are used as a recreation ground for people from the nearby towns. However, large areas are undisturbed and the wood holds an interesting breeding bird community and attracts some migrants on passage.

Strategy   The site is worth visiting at any time of year but spring is undoubtedly the best time. The site has very good populations of Red-necked Nightjar, Bee-eater, Woodlark and Woodchat Shrike. Sundays should be avoided as many picnickers go to the Pinar for the day. The first hour after sunset is best for nocturnal birds. The site can be visited on its own for a half day or less or it can be incorporated in an itinerary including other sites close by, such as Almoraima and Guadiaro.

The Pinar del Rey is situated 3km north of the town of San Roque and access is possible from several directions. If travelling from Almoraima, take a left turning off the C 3331 just past Km86, before reaching the Estación de San Roque. This country road goes over a bridge and then continues through a mixture of woodland and scrub which are rich in birds during the breeding season. After about 3km the road turns right sharply; just here, on the left, there is a track leading to a gate which is kept open. This is the entrance to the Pinar. Follow the main track and stop at intervals to wander into the woods. Alternatively, the road can be followed from the opposite direction, by passing though San Roque and following the road that leads out of the town northwards until the Pinar is reached after 3km or by taking a turning off the N 340 at Km118 which also links up with this road.

Red-necked Nightjar

Birds   Interesting breeding species include Booted Eagle, Scops and Tawny Owls, Red-necked Nightjar, Bee-eater, Wryneck, Woodlark, Black-eared Wheatear (edges of the wood), Cetti's, Orphean and Bonelli's Warblers, Firecrest, Crested Tit, Short-toed Treecreeper, Golden Oriole, Woodchat Shrike, Spotless Starling, Hawfinch and Cirl Bunting.

Playa de Los Lances

This 4km long sandy beach at the western end of the Strait of
Gibraltar, faces the Atlantic Ocean. The Jara and Vega rivers join
together and flow into the sea at the eastern end of the beach. Tidal
pools form on the beach at regular intervals. The landward side of the
beach has small dunes with typical sandy vegetation. Behind the dunes
there is a belt of Stone Pines. Los Lances is ideally located on the western coastal approach to Tarifa. It therefore receives a wide variety of
migratory species which settle on the beach and adjacent habitats to rest
and feed. It is also well placed for observing active diurnal migration.

Strategy

This site is worth a visit at any time of the year although it is at its
best during the migration periods. Early morning is the best time to
visit, before the beach is disturbed by bathers and wind surfers.
Easterly winds are the best for observing active migration. Sea-watching can be rewarding at any time.

Key

Pines

Roads

Rivers

Tracks

Mountain Peaks

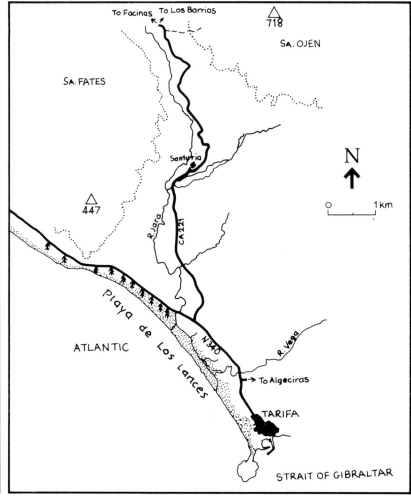

Playa de los Lances

Follow the N 340 west from Tarifa towards Vejer and take a left turning towards the beach, through the coastal pine wood, after crossing over the Jara River. Drive through the narrow belt of pines and park by the beach. A walk along the entire length of the beach is recommended.

Rufous Bush Robin

Birds

Breeding species include Kentish Plover, Short-toed Lark, Melodious Warbler, Woodchat Shrike and Serin.

Typical migrants and wintering birds include Cory's and Balearic Shearwaters, Gannet, Purple Heron, Black and White Storks, Spoonbill, Greater Flamingo, Garganey, Common Scoter, Honey Buzzard, Black Kite, Egyptian and Griffon Vultures, Short-toed Eagle, Marsh, Hen and Montagu's Harriers, Booted Eagle, Osprey, Lesser Kestrel, Hobby, Quail, Water Rail, Black-winged Stilt, Stone-curlew, Little Ringed Plover, Knot, Sanderling, Little Stint, Curlew Sandpiper, Ruff, Black-tailed and Bar-tailed Godwits, Whimbrel, Spotted Redshank, Greenshank, Green and Wood Sandpiper, Grey Phalarope, Pomarine, Arctic and Great Skuas, Mediterranean, Little and Audouin's Gulls, Kittiwake, Gull-billed, Royal (occasional), Lesser Crested (occasional), Whiskered and Black Terns, Razorbill, Puffin, Scops Owl, Nightjar, Red-necked Nightjar, Alpine Swift, Bee-eater, Hoopoe, Wryneck, Short-toed Lark, Woodlark, Crag Martin, Red-rumped Swallow, Tawny Pipit, Rufous Bush Robin, Black Redstart, Redstart, Whinchat, Black-eared Wheatear, Olivaceous, Melodious, Spectacled, Subalpine, Orphean and Bonelli's Warblers, Pied Flycatcher, Golden Oriole, Woodchat Shrike, Serin, Siskin and Ortolan Bunting.

Other Wildlife

Spiny-footed Lizard, Ocellated Lizard, Otter.

Jara Valley

The Jara River rises in the Sierra de Ojén and follows a short course to the sea along a valley carved out between the Sierra de Ojén in the east and the Sierra de Fates in the west. It reaches the sea at Los Lances beach, west of Tarifa. The valley of the Jara can be divided into two portions. The upper reaches are dominated by a mixture of Cork Oak woodland, maquis dominated by *Pistacia lentiscus,* and riverine vegetation with *Rubus ulmifolius* and White Poplar. The lower portions are characterised by a narrow belt of riverside vegetation dominated by Oleander and Rubus with the surrounding flat or hilly countryside devoted to pasture.

Strategy   The Jara Valley is excellent at any time of year. In spring and autumn it is ideally situated to channel streams of migrants following the western end of the Strait of Gibraltar. In late July and early August large flocks of White Storks and Black Kites rest in the valley before crossing the Strait and many roost in the woods on the edge of the valley. In strong easterlies when the sea crossing is not possible, thousands may be seen grounded.

Key

Pines

Roads

Rivers

Tracks

Mountain Peaks

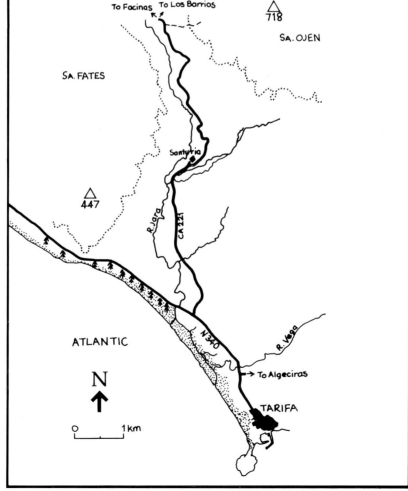

Follow the N 340 from Tarifa to Vejer and take a right turning at Km80. A narrow country road which is in good condition (CA 221) follows the base of the Jara Valley northwards. A number of footpaths and tracks lead away from the road towards the river or the surrounding countryside.

Birds   Interesting breeding species, including species which feed in the valley but breed in the nearby sierras, are: Egyptian and Griffon Vultures, Short-toed Eagle, Booted and Bonelli's Eagles, Peregrine, Scops and Eagle Owls, Red-necked Nightjar, White-rumped Swift, Bee-eater, Calandra Lark, Red-rumped Swallow, Tawny Pipit, Rufous Bush Robin, Black-eared Wheatear, Cetti's, Olivaceous, Melodious and Bonelli's Warblers, Crested Tit, Short-toed Treecreeper, Woodchat Shrike, Raven, Spotless Starling, Serin, Hawfinch and Cirl Bunting.

Typical migrants and/or winter visitors are: Black and White Storks, Honey Buzzard, Black and Red Kites, Egyptian and Griffon Vultures, Short-toed Eagle, Marsh, Hen and Montagu's Harriers, Booted Eagle, Osprey, Lesser Kestrel, Merlin, Hobby, Quail, Little Bustard, Stone-curlew, Golden Plover, Scops Owl, Nightjar, Red-necked Nightjar, Alpine Swift, Bee-eater, Roller, Hoopoe, Wryneck, Short-toed Lark, Woodlark, Crag Martin, Red-rumped Swallow, Tawny Pipit, Rufous Bush Robin, Black Redstart, Redstart, Whinchat, Black-eared Wheatear, Olivaceous, Melodious, Dartford, Spectacled, Subalpine, Orphean and Bonelli's Warblers, Pied Flycatcher, Golden Oriole, Great Grey and Woodchat Shrikes, Serin, Siskin, Hawfinch and Ortolan Bunting.

Zahara de los Atunes and Cape Trafalgar

The coastal belt of the south-western corner of Spain is an attractive area which holds some interesting breeding birds and which attracts many migrants moving along the Atlantic coast, between Spain and Morocco, or between the Strait of Gibraltar and the Atlantic. The apex is at Cape Trafalgar where the south-east running coastline takes a sharp turn to almost due east. The Cape itself is on a sand spit which juts into the Atlantic and sea-watching can be interesting from the small cliffs by the lighthouse. There is also active diurnal migration here, particularly during easterly winds. The sandy beaches around the cape and the shallow lagoons that form there attract waders, gulls and terns and the Stone Pine woods of the area hold migrants in spring. The Barbate pine wood, east of Cape Trafalgar, is large and has a few interesting breeding birds; it attracts migrants on passage. The sandstone cliff between Barbate and Cape Trafalgar has a large egret colony (over 2,600 pairs of Cattle Egrets and 50 pairs of Little Egrets). Formerly, it held breeding Shags and Ospreys. There are three estuaries in the area. The largest is the Barbate, just east of the town, which has a large complex of abandoned salt pans. Just west of Zahara, three streams (Arroyos de la Zarzuela, Acebuchal and Candalar) open into the sea in a small estuary

Just south of Conil, the small estuary of the Rio Salado opens into the Atlantic. The three estuaries, especially the Barbate, attract waders and seabirds on passage and in winter.

Key

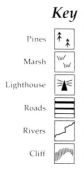

Pines

Marsh

Lighthouse

Roads

Rivers

Cliff

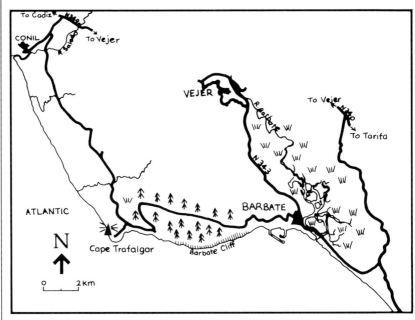

Strategy   The itinerary described below can be followed during a quick half-day visit or as a leisurely full day outing. Spring is best since there is a good chance of seeing grounded migrants and, during easterly winds, raptors which have drifted from the Strait area. Autumn is also a good time. In winter, the estuaries hold some waders but there are fewer birds in the pine woods. The cliff between Barbate and Trafalgar is of little interest outside the breeding season. Sea-watching from Cape Trafalgar is possible the year round but best during the passage periods.

From the Venta de Retín on the N 340 (see La Janda, p 36), a country road leads south-west for 10.5km to the small coastal village of Zahara de los Atunes. Follow the coastal road past Zahara in a westerly direction towards Barbate. The small river which passes Zahara and the small estuary should be checked for waders and other migrants. Between Zahara and Barbate the road crosses a military area so don't stop there. Nine kilometres after Zahara the salt pans and estuary of the Barbate appear on either side of the road. Stop the car in a lay-by and search the area for waders and seabirds. It is possible to walk into the salt pans but the marsh upstream of the River Barbate is fairly inaccessible. From here continue west through the town of Barbate towards Los Caños de Meca. On leaving Barbate the road climbs up a hill and crosses the impressive Barbate pine wood for 11km. It is possible to stop at many points here and wander into the woods in search of birds. After 5km from Barbate, stop and walk south through the pine woods towards the sea. Eventually the top of the Tajo de Barbate is reached from where the cliff nesting birds can be observed.

The road carries on west to the coastal resort village of Caños de Meca and, from here, west towards Cape Trafalgar. From Cape Trafalgar a country road leads north-west towards Conil de la Frontera, following the coast through cultivated fields which, in spring, attract Montagu's Harriers. Before reaching Conil, the small estuary of the Rio Salado is reached. From Conil, the CA 213 runs north-east for 2.5km to rejoin the N 340 north-west of Vejer.

Birds

Most of the migrants which cross the Strait can appear in the area (see species lists for Gibraltar, Algeciras sierras, Tarifa, La Janda and Guadiaro, Palmones and Guadarranque estuaries). This is a good area for passage Rollers. Interesting breeding species include Griffon and Egyptian Vultures (from nearby sierras), Montagu's Harrier, Booted Eagle, Peregrine, Stone-curlew, Kentish Plover, Scops Owl, Long-eared Owl (scarce), Red-necked Nightjar, Woodlark, Black-eared Wheatear, Melodious, Dartford and Bonelli's Warblers, Firecrest, Short-toed Treecreeper, Woodchat Shrike and Cirl Bunting. In winter, Caspian Terns are regular in the Barbate estuary.

La Janda

La Janda was the famous freshwater lagoon, north-west of the Strait, where the last southern European Cranes nested. The lagoon was progressively drained during the 1950's and the rich breeding communities of the lagoon, which included Squacco Heron, Glossy Ibis, Ferruginous Duck, Purple Gallinule, Crested Coot and African Marsh Owl were lost.

Despite this great loss, La Janda remains one of southern Spain's best birdwatching sites although few visiting birdwatchers ever go there. It is at its best in winter and spring when large areas become flooded and the large expanse of flat, low-lying land, surrounded by sierras in all directions, makes La Janda an attractive hunting ground for many raptors.

Strategy

Any time of year is good for La Janda. Winter is excellent for waders, cranes and raptors. Red Kites and Hen Harriers regularly winter here and it is one of the few regular wintering sites for Booted Eagles which are essentially summer visitors to Europe. In recent winters, scarce raptors have been regularly seen, including Spanish Imperial Eagle, Black-shouldered Kite and Black Vulture. Cranes also winter in large numbers with counts of 2,000 in recent years.

La Janda is also a focal point for migrants travelling towards the Strait of Gibraltar along the Atlantic side of Iberia. In late July and August, thousands of White Storks and Black Kites often rest in the dry fields awaiting favourable winds for crossing the Strait. Impressive roosts can be seen with hundreds of storks and more than 300 Montagu's Harriers have been recorded at a single roost in mid-August During passage periods most species which cross the Strait are likely to be seen in the area. This is a good site for migrant Rollers and if the

Key

Oaks

Roads

Rivers

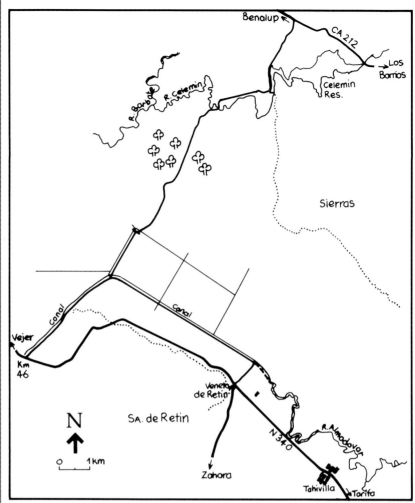

fields are flooded, passage waders can be seen in large numbers. North African rarities sometimes occur – Cream-coloured Courser and Little Swift have been seen in recent years.

A spring visit will benefit from the presence of both migrants and breeding species. La Janda is the only breeding site in the province of Cádiz for Great Bustard, but fewer than 10 individuals now remain.

The fields near Tahivilla (Km 59 on the N 340) are the most likely site for Great Bustard. The fields around the country road to Zahara de los Atunes, between the Venta de Retín and the village of La Zarzuela, sometimes attract Great Bustards as well as other typical open country species like Montagu's Harrier and Calandra Lark.

A complete day can easily be spent in La Janda. Early morning (first light) is best for shy species such as the bustards which usually disappear by mid-morning.

La Janda is situated north-east of Vejer. There are three main points of access. The first two are from the N 340. Travelling north-west from Tarifa, take the right turning (between Km 54 and Km 55) down a track by the Venta de Retín opposite the turning to Zahara de los Atunes. The track leads to a T-junction after about 1km. Take the left turning through a red gate which is usually open. If it is closed, open it and pass but make sure to close it again. Before continuing, the fields around here should be checked thoroughly with telescope as Little Bustards are frequently seen. Watch out for the occasional Great Bustard.

The track then continues west for 5km. In dry weather a car can go the whole way but in wet conditions a 4-wheel drive vehicle is necessary from about the mid-point. Stop at intervals along the track and check the surrounding fields, which in the wet season often have pools. There is a large drainage canal with aquatic vegetation which follows the track and which is good for aquatic species. The track eventually reaches a country road which can be approached from another point on the N 340. Those unable to follow the track all the way should turn back and rejoin the N 340 at Venta de Retín and from there continue northwest until Km46. Here a road to the right passes through a similar open red gate to the one at Venta de Retín. This country road travels northeast along another drainage canal and stops along here are recommended. Eventually the road meets the western end of the track from Venta de Retín by a bridge which crosses the main drainage canal. Cranes often feed in the fields around this bridge in winter. The road continues north for 10km, passing cultivated fields, then climbing towards a plateau, past a farm, which is lightly wooded with Cork Oaks, descending towards other cultivated fields, finally passing a dam and the Celemín Reservoir before reaching the CA 212 which runs from Benalup to Los Barrios. This junction is the third access point to La Janda. The Celemín Reservoir, although not usually very good, is worth checking, especially during migration periods.

Little Bustard

Birds   Interesting breeding species, including those from nearby sierras which hunt over the area, are: Little Bittern, White Stork, Egyptian and Griffon Vultures, Short-toed Eagle, Montagu's Harrier, Booted and Bonelli's Eagles, Quail, Pheasant (introduced), Baillon's Crake (possible Little and Great Bustards, Stone-curlew, Collared Pratincole, Scops and Eagle Owls, Nightjar, Bee-eater, Calandra and Short-toed Larks, Black-eared Wheatear, Cetti's and Great Reed Warblers, Raven and Tree Sparrow.

In winter the following species are of interest, the presence of some being determined by water levels: White Stork, Black-shouldered (scarce) and Red Kites, Griffon Vulture, Black Vulture (scarce), Marsh and Hen Harriers, Spanish Imperial (scarce), Golden (scarce), Booted and Bonelli's Eagles, Merlin, Peregrine, Pheasant, Water Rail, Crane, Little and Great Bustards, Stone-curlew, Kentish and Golden Plovers, Sanderling, Ruff, Black-tailed Godwit, Green Sandpiper, Mediterranean and Little Gulls, Eagle and Short-eared Owls, Wryneck, Calandra Lark, Crag Martin, Water Pipit, Cetti's Warbler, Bluethroat, Penduline Tit, Great Grey Shrike and Spanish and Tree Sparrows.

Other Wildlife   Montpellier, Ladder and Horshoe Whip Snakes, Ocellated Lizard, Stripe-necked Terrapin, Egyptian Mongoose, Weasel, Garden Dormouse, Red Fox and Genet.

Benalup Woodland

North-west of La Janda, the Barbate River flows through lowland Cork Oak woodland between the towns of Alcalá de los Gazules and Benalup de Sidonia. Together with Almoraima, these woods are among the most extensive and undisturbed lowland woods in the area and they abound with breeding species in spring including a large population of Hoopoes. In winter, they hold the typical wintering woodland birds of the area and at other times there is always the chance of a passing migrant - Roller, for example, is regularly seen here.

Strategy   This site is best visited in spring for breeding species although visits at any time of year can add a range of woodland species to the Janda list. In spring especially, morning or evening visits are best as bird activity is then at its highest.

If leaving La Janda via the Celemín Reservoir, turn left at the main road (CA 212) which runs towards the town of Benalup. On reaching the edge of the town, take a right turning marked Alcalá de los Gazules. This country road runs for 18km through the Cork Oak woodland. Stop at intervals and watch from the roadside. After 10km, there is a firm track on the right which leads to a construction site for a dam. This track can be followed in a car and runs through woodland which is good for birds.

As an alternative, on joining the CA 212 at Celemín, take the track immediately opposite. This track, which can also be followed by car, runs through farmland. Although not as good as the other road, it can complement a visit to the woodland site by adding open ground species such as White Stork and Montagu's Harrier.

Birds   Interesting breeding birds include White Stork, Black Kite, Egyptian and Griffon Vultures (from nearby sierras), Short-toed Eagle, Goshawk, Booted Eagle, Scops Owl, Bee-eater, Hoopoe, Green Woodpecker,

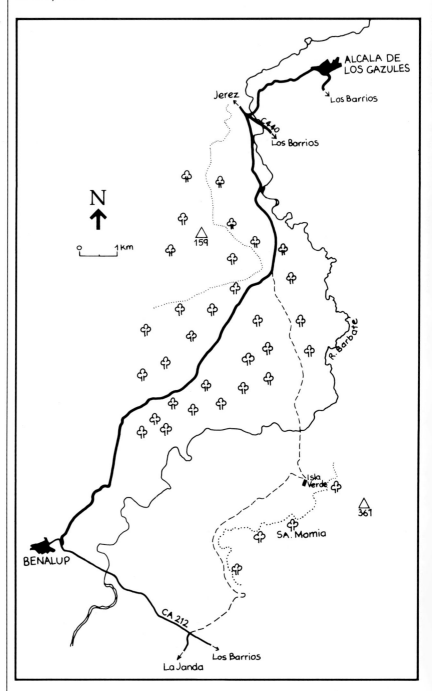

Woodlark, Red-rumped Swallow, Black-eared Wheatear, Cetti's, Melodious and Bonelli's Warblers, Firecrest, Short-toed Treecreeper, Golden Oriole, Woodchat Shrike, Spotless Starling, Serin, Hawfinch and Cirl Bunting.

Other Wildlife   Red Fox, Weasel, Western Polecat, Egyptian Mongoose, Genet, Ocellated Lizard and Ladder Snake.

Alcalá de los Gazules - Llanos del Valle

This is an area of rolling hills with large amounts of pasture and cultivation. The higher parts, in particular, are covered in open woodland dominated by Olive, Holm Oak and some Cork Oak. Other parts have mainly Olive and Lentisc scrub while Oleander scrub predominates along river banks, with Prickly Pear hedges obvious in certain areas.

The Sierras de La Sal (501m) and de Las Cabras (683m) are limestone outcrops among the sandstone rocks. Small limestone outcrops like Peña Arpada stand out in the farmland between Alcalá and Paterna de Ribera.

Strategy

The area can be approached from Alcalá de los Gazules in the south, from Puerto Galiz in the south-east, from Arcos de la Frontera or Jerez de la Frontera in the north-west or from Paterna de Ribera in the west.

Key

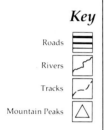

Roads

Rivers

Tracks

Mountain Peaks

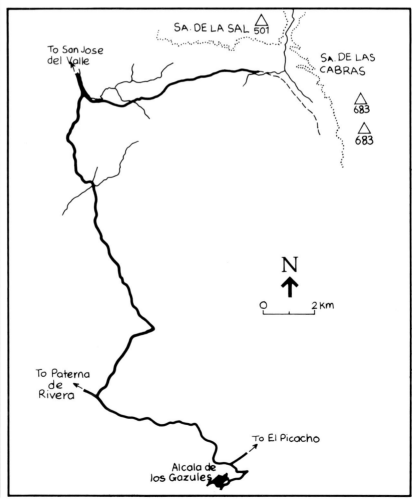

From Alcalá, take the road signposted 'El Picacho' and after 2.5km take the road to Paterna de Ribera. After 10km take a right turning to San José del Valle and another right turn after 25.4km to Llanos del Valle. This last road is in good condition for about 8km but then deteriorates into a poor track, suitable only for four-wheel drive vehicles. The track is worth following for at least one kilometre.

The best time to visit is from April to August although it is interesting at other times of year. Mornings and evenings are best when small birds are active and raptors are not flying too high.

Montagu's Harrier

Birds   Breeding species include: Egyptian and Griffon Vultures, Short-toed Eagle, Montagu's Harrier, Booted and Bonelli's Eagles, Lesser Kestrel, Peregrine, Scops Owl, Red-necked Nightjar (best seen at dusk from May to August), Hoopoe, Woodlark, Red-rumped Swallow, Rufous Bush Robin (in dry river-bed scrub), Black-eared Wheatear, Cetti's and Melodious Warblers, Woodchat Shrike, Spotless Starling and Spanish and Tree Sparrows. Among the many migrants, Rollers are seen in small numbers in late April/May and late August/early September.

Other Wildlife   Ocellated Lizard.

Cádiz Lagoons

There are about a dozen excellent freshwater lagoons in the province of Cádiz. Some are on private land, others are of difficult access. The largest and most accessible, and two others which are close to it, are described. Between them they hold all the interesting freshwater birds of southern Spain. The lagoons are typically shallow freshwater lakes with inflow provided by one or more streams. The inflow is usually insufficient to keep the water level constant in summer and some lagoons dry completely at that season, especially after a dry winter. The changing water levels are important in the lagoon at any particular time. After a dry period a lagoon may have a large shoreline which will attract waders. After heavy rains the lagoons fill and become too deep even for flamingoes but attract ducks in great numbers. Most lagoons are characterised by a vegetated margin of reed and associated fresh-water vegetation where most species breed. The countryside surround-ing the lagoons is usually cultivated. Most lagoons are now protected.

   The three lagoons described are Laguna de Medina, Laguna Salada and Laguna de las Pachecas. Medina is the largest, being over 1km long and about $\frac{1}{2}$ km wide. Salada is smaller and Pachecas is very small, completely covered in reeds and is close to some houses. All, except Pachecas, are Natural Reserves.

Key

Roads

Tracks

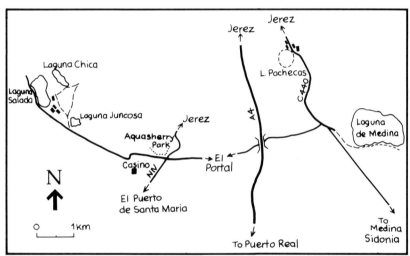

Strategy   The lagoons are interesting at any time of year. In spring many breeding species can be seen. In winter the variety of duck species increases and during passage many waders and wildfowl can be found.

   All three lagoons can be visited in a single trip which should be in the morning to avoid heat haze which is a problem for telescope users, although less so in winter. On a day trip from Gibraltar another site such as La Janda or one of the sierras can be included on the return lap.

   Follow the C 440 north from Alcalá de los Gazules towards Jerez de la Frontera. After about 45km (5km before reaching the intersection with the A 4 Motorway) there is a sign indicating a right hand detour to the Laguna de Medina. The lagoon is visible from the main road. As the lagoon is approached from the C 440, a white building on the left of the road with the words 'El Calvario' painted on its side, is a landmark to look out for. Once the track to the lagoon has been taken, follow it until a gate is reached. Leave the car here and follow a track which leads along the southern bank of the lagoon in an easterly direction. There are several vantage points along this track and a telescope is advantageous.

The Laguna de las Pachecas can be easily reached from the Laguna de Medina by returning to the C 440 and turning right in the direction of Jerez. Look out for a group of white buildings on the left after about 1km. Take a track at this point which goes round a reed-covered lagoon. Stop at the northern, elevated, end of the lagoon from where the whole area can be watched. Purple Gallinules, Little Bitterns and Savi's Warblers are frequently seen from here.

Return to the C 440 and turn right, in the direction of Medina, but turn right again towards El Portal before reaching Medina. On reaching El Portal, turn left past the railway line and continue on the CA 201 until the main N IV is reached. Cross the N IV and, immediately opposite, follow a road which passes the Aquasherry Park on the right and, later, a Casino on the left. After 3km from the N IV take a track on the right which leads to some isolated houses on cultivated land. Park here and walk along the track for about 100m to the shore of the Laguna Salada from where vantage points can be selected. From where the car is parked another track leads to the right, going to a small lagoon (Laguna Chica). This lagoon, which attracts waders on passage, should also be checked. In winter, a third, reed-covered, lagoon forms at the junction of the track with the road.

Birds

Interesting breeding species include: Black-necked Grebe, Little Bittern, Purple Heron, Red-crested Pochard, White-headed Duck, Marsh and Montagu's Harriers, Purple Gallinule, Crested Coot (rare), Black-winged Stilt, Collared Pratincole, Bee-eater, Cetti's, Savi's, Great Reed and Melodious Warblers and Woodchat Shrike.

On passage and/or in winter the following additional species are of interest: Night Heron (scarce), White Stork, Greater Flamingo, Greylag Goose, Shelduck, Pintail, Garganey, Marbled Duck (irregular but sometimes in flocks), Honey Buzzard, Black Kite, Red Kite (regular in winter), Egyptian and Griffon Vultures, Hen Harrier (regular in winter), Osprey, Lesser Kestrel, Water Rail, Baillon's Crake (may breed), Avocet, Stone-curlew, Little Ringed, Kentish and Golden Plovers, Knot, Sanderling, Little Stint, Temminck's Stint (scarce), Curlew Sandpiper, Ruff, Black-tailed and Bar-tailed Godwits, Spotted Redshank, Marsh Sandpiper (scarce), Greenshank, Green and Wood Sandpipers, Mediterranean Gull, Gull-billed, Whiskered, Black and White-winged Black (rare) Terns, Short-eared Owl, Hoopoe, Wryneck, Red-rumped Swallow, Pied Flycatcher, Penduline Tit (regular in winter) and Great Grey Shrike.

White-headed Duck

Guadalquivir – south-east bank

The Guadalquivir is the largest river in southern Spain. As it winds its way towards the Atlantic past Sevilla, it deposits fluvial sediments which add to the existing marshes which have become known by the Spanish name of 'Marismas'. The west bank, which includes the Parque Nacional de Doñana, is the most visited and best known, but the east bank is also very interesting and rich in birdlife. For the visitor staying in Gibraltar or the province of Cádiz, the east bank has the advantage that it can be reached without having to drive to Sevilla and be comfortably visited in a day.

The east bank consists of large areas of marsh some parts of which have been drained for agriculture. Towards the estuary, near the town of Bonanza, there are salt pans which attract large numbers of water-birds and waders. The Stone Pine wood at Monte Algaida between the salt pans and the marsh has most of the typical species of the Doñana pine woods.

Key

Pines

Marsh

Roads

Rivers

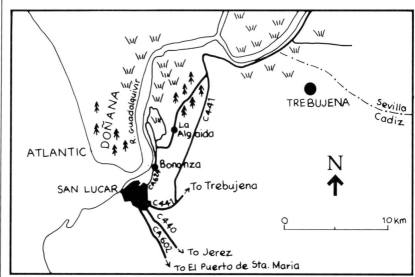

Accommodation

Accommodation is available in some of the nearby towns although Sanlucar is not recommended. Jerez is close and has a good range of hotels and hostels (1-star to 5-star. Price range for double room 1,200-16,590 pesetas per night). The closest camping sites are in Rota and El Puerto de Santa Maria.

Strategy

The area is excellent at any time of year: passage waders and wildfowl occur in large numbers, in winter Greylag Geese can be seen alongside Greater Flamingoes and in the summer many typical marsh species breed. The pine wood attracts many raptors which nest in the pines and hunt over the marsh, especially Black and Red Kites, while Azure-winged Magpie and Great Spotted Cuckoo breed there. Allow a whole day for the area. From Gibraltar, the site is about $2\frac{1}{2}$ hours away by car so a day trip is possible.

The best time for observing birds in the marsh is morning and evening when birds are most active and heat haze does not limit the use of a telescope. The salt pans and the pine wood are best at these times of day, especially in spring and summer, but are productive at any time of day. Large concentrations of birds may be expected at most times.

From the N IV take either the CA 602 or the C 440 towards Sanlucar de Barrameda. If Laguna Salada has been visited, continue from there westwards along the country road which joins the CA 613. Turn left here and within 100m the junction with the CA 602 will be reached. At Sanlucar, travel towards the city centre. At the first traffic lights there is a sign indicating a right hand turn to Bonanza. This is a bad road which should be avoided. Continue instead towards the city centre and at the second traffic lights turn right where there is another sign to Bonanza. Follow the street until a small bullring is passed on the right. Turn left onto a main street and right on reaching the street. Continue along this road, passing close to the Guadalquivir River.

On reaching the obvious salt works there is a building on the left of the road with the word 'Apromasa' painted on its side. Just before this building, on the right of the road, there is a small white house with bars on the windows. These are the salt company's offices and permission to enter the salt pans must be obtained here - this is normally given readily to birdwatchers. The office is open on weekdays from 9am to 1pm and from 3pm to 6pm. If visiting the site on a weekend a permit should be obtained in advance by writing to APROMASA, Apartado de Correos, 111, Sanlucar de Barrameda, Cádiz [telephone: (956) 36 07 19].

The entrance to the salt pans is via a track on the left about 100m further on from the office. The track is usually firm and can take a car. Drive past the gate along the central track and stop at intervals to observe the birds on the salt pans. At the end of the track there is a fork. Park here. Walk along the right fork to reach other salt pans which are usually interesting, attracting Spoonbills and Greater Flamingoes. The left track leads to the river bank where there are usually terns flying up and down the river. From here it is possible to wander into the Bonanza side of the marsh. Cars should not be driven here as the track is bad in places.

Returning to the main road, turn left and follow the road (CA 624) which leads to a Stone Pine wood. Follow the track through the centre of the wood and stop at intervals to observe the woodland birds. At the end of the wood (after about 5km), the track joins a road. Turn left and follow the road which goes through a marsh beside a drainage canal. Stop along the road and scan the marsh. 9km from the edge of the pine wood, there is a sign indicating the provincial boundary with Sevilla. Continue for another kilometre and a half and turn right down a track which enters the marsh. 2.5km from the road there is a grassy patch which in spring usually holds Pin-tailed Sandgrouse and Stone-curlew. Return along the track and back towards the pine wood. There are one or two more tracks into the marsh which can be explored. The best spots for scanning vary with season and according to water levels. The marsh on the opposite bank of the river should also be checked.

Birds   In spring/summer interesting species (including breeding species) are: Little Bittern, Night Heron, Squacco Heron (scarce), Purple Heron, Black Stork (scarce), White Stork, Spoonbill, Greater Flamingo, Garganey, Marbled Duck (erratic), Black and Red Kites, Egyptian Vulture (occasional), Griffon Vulture (few), Marsh and Montagu's Harriers, Spanish Imperial Eagle (may wander from Doñana), Hobby, Quail, Black-winged Stilt, Avocet, Stone-curlew, Collared Pratincole, Little Ringed and Kentish Plovers, Knot, Sanderling, Little Stint, Curlew Sandpiper, Ruff, Black-tailed and Bar-tailed Godwits, Whimbrel, Spotted Redshank, Marsh Sandpiper (scarce), Greenshank, Green and Wood Sandpipers, Grey Phalarope, Slender-billed Gull, Gull-billed, Caspian, Whiskered and Black Terns, Pin-tailed Sandgrouse, Great Spotted Cuckoo, Scops Owl, Red-necked Nightjar, Bee-eater, Roller, Hoopoe, Wryneck, Calandra, Short-toed and Lesser Short-toed Larks, Woodlark, Cetti's, Great Reed, Melodious and Spectacled Warblers, Short-toed Treecreeper, Woodchat Shrike, Azure-winged Magpie, Raven, Spotless Starling and Tree Sparrow.

In winter, the following species are of interest in addition to some noted above: Greylag Goose, Shelduck, Pintail, Red-crested Pochard, Hen Harrier, Osprey, Merlin, Peregrine, Water Rail, Little Bustard, Golden Plover, Short-eared Owl, Penduline Tit and Great Grey Shrike.

Other Wildlife   Ocellated Lizard, Chameleon.

Marismas del Guadalquivir/ Doñana National Park

To the west of the Rio Guadalquivir, south of Sevilla, is a huge marsh, a section of which has remained relatively undisturbed as a result of its incorporation within the Parque Nacional de Doñana. Towards the coast a wide belt of sand dunes, heavily wooded with Stone Pines, separates the marshes from the sandy beaches and the Atlantic Ocean. It is this unique combination of lowland woodland and marsh spreading over a large area of south-west Spain which has made the estuary of the Guadalquivir and the Coto Doñana in particular, a world famous biological reserve. The National Park covers 50,720 hectares. The surrounding area, covering an additional 56,930 hectares, is a Natural Park.

The area is important for its breeding species which include some very scarce waterbirds and a small population of Spanish Imperial Eagles, for its wintering populations of Greylag Geese and ducks, and as a transit area for migrants moving between Europe and Africa along the Atlantic flyways.

Accommodation   There are 2- and 3-star hotels in Matalascañas and 1-star accommodation is available at El Rocio. There is a 3-star Parador Nacional 23km west of Matalascañas on the C 442 to Huelva.

Key

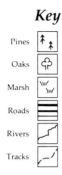

Pines
Oaks
Marsh
Roads
Rivers
Tracks

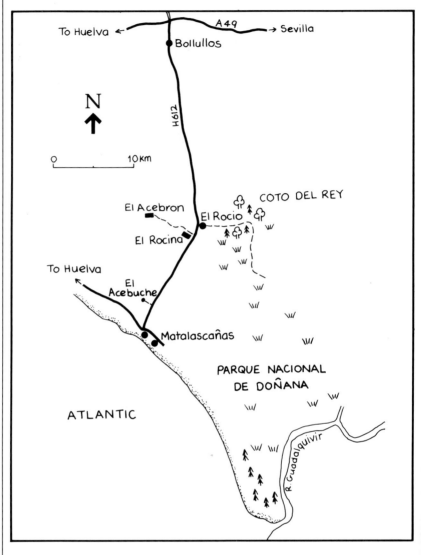

Strategy   Doñana is excellent at any time of year but the best times for a visit are April-June for breeding birds and migrants and December-February for wintering birds. The summer, with the drought, is the quietest period.

Those visiting the park for the first time should spend half a day on the land rover tour which can be combined with a visit to El Acebuche and La Rocina and watching from the hides. A second day can be usefully spent in the Coto del Rey and a third day by visiting the Odiel Estuary near Huelva (described under brief notes).

Those with limited time can arrange a tour of the park from Sanlucar de Barrameda which has the advantage of avoiding the long drive to Sevilla. Land Rovers collect passengers from the park's interpretation centre in Sanlucar and transport them for a tour of the park by barge across the Guadalquivir. These tours are run by the same cooperative that runs the tours from El Acebuche.

To get to the west bank of the Guadalquivir it is first necessary to go to Sevilla. If coming from the Strait of Gibraltar area, then it is best to approach Sevilla via the C 440 from Los Barrios to Jerez de la Frontera. This road joins the A4 (E5) motorway before Jerez and continues to Dos Hermanas on the southern edge of Sevilla. Follow the signs to Huelva by keeping west, thus avoiding Sevilla altogether. This should lead to the A 49 (E01) which runs west out of Sevilla. Continue towards Huelva for 51km to Bollullos par del Condado and turn left here on the H 612 towards Almonte. This road leads south, past Almonte, to El Rocio (27km) and Matalascañas (43km).

Between El Rocio and Matalascañas are the Park's interpretation centres. The two most worth visiting are La Rocina and El Acebuche. From La Rocina, a track leads west to the Palacio del Acebrón which has planned walks, an exhibition centre and a museum about Man and the Marismas. El Acebuche lies close to Matalascañas and has an interesting museum and a small lagoon with hides where it is possible to observe some of the typical waterbirds of the area. From El Acebuche it is possible to arrange for a tour of the Park in a land rover. Access to the park is restricted to these tours. The tour, which takes several hours, is a good introduction and offers opportunities to observe some interesting species, although during the breeding season some areas are not visited. At certain times of year the tours are booked well in advance and it is advisable to arrange for a trip well before the visit by telephone [El Acebuche (955) 430432 or 406722 or La Rocina (955) 406140] or by writing to Coop. Marismas del Rocio, Parque Nacional Doñana, Centro del Acebuche, 21760 Matalascañas, Huelva.

Those wanting to explore for themselves should visit the Coto del Rey which lies outside the main park area but which holds most of the species likely to be met in the area. Take the track into El Rocio which passes the church on the left. Take the second left hand turning down a street within El Rocio which leads to the edge of the town. Follow the track, over a small bridge. This leads to an area of oak and pine woodland which marks the start of the Coto del Rey. The track continues for some time within the wood and it is possible to stop and wander in many places. Eventually it reaches the marisma. How far into the marsh one can go depends on time of year and the condition of the track and paths which is governed by season and weather. A whole day can easily be spent exploring this area.

Just outside El Rocio a bridge on the H 621 to Matalascañas crosses the Arroyo de la Rocina. This is a good site to scan the reed beds for Savi's Warbler which can also be observed from the hides at the nearby interpretation centre.

From Matalascañas, the C 442 leads west to Huelva passing coastal Stone Pine woods which are accessible and can also be explored for birds.

Birds   It is difficult to highlight interesting birds in such a rich area. Among the breeding species the following are important: Purple, Night and Squacco Herons, Little Bittern, Spoonbill, Greater Flamingo, Marbled Duck, Black and Red Kites, Spanish Imperial, Booted and Short-toed

Eagles, Marsh Harrier, Peregrine, Baillon's Crake, Purple Gallinule, Crested Coot, Collared Pratincole, Black-winged Stilt, Avocet, Kentish Plover, Slender-billed Gull, Gull-billed and Whiskered Terns, Great Spotted Cuckoo, Scops Owl, Red-necked Nightjar, Short-toed, Lesser Short-toed and Thekla Larks, Savi's and Spectacled Warblers and Azure-winged Magpie.

Among the main wintering species are: Greylag Goose (about 70,000), Shelduck, Pintail, Red-crested Pochard and Black-tailed Godwit.

Purple Gallinule

Other Wildlife   Western Hedgehog, Greater Horshoe Bat, Large Mouse-eared Bat, Serotine, Mediterranean Pine Vole, Red Fox, Badger, Weasel, Western Polecat, Otter, Egyptian Mongoose, Genet, Wild Cat, Lynx, Wild Boar, Red Deer, Fallow Deer, Parsley Frog, Western Spadefoot, Natterjack, Painted Frog, Iberian Midwife Toad, Tree Frog, Marbled Newt, Bosca's Newt, Sharp-ribbed Salamander, Bedriaga's Skink, Three-toed Skink, Spiny-footed Lizard, Large Psammodromus, Spanish Psammodromus, Iberian Wall Lizard, Moorish Gecko, Amphisbaenian, Ocellated Lizard, Montpellier Snake, Ladder Snake, Viperine Snake, Smooth Snake, Lataste's Viper, Spur-thighed Tortoise, European Pond Terrapin, Stripe-necked Terrapin.

Serranía de Ronda

The Serranía de Ronda is a large mass of high mountains west of the Sierra Nevada and north-east of the Strait of Gibraltar. The precise boundaries of the Serranía differ according to different authorities. Here, the widest interpretation is used from the sierras around Grazalema in the north-west to those near Marbella in the south-east. Also included is the large mass of volcanic peridotite known as the Sierra Bermeja which is a coastal sierra behind the town of Estepona and which has a maximum altitude of 1450m. The highest point in the Serranía is Torrecilla (1919m).

Accommodation

Accommodation is available in hotels and hostels in many of the towns in and around the Serranía (Ronda, Grazalema, Ubrique, Gaucín, Estepona, San Pedro de Alcantara).

Strategy

The Serranía has the greatest bird diversity in spring but interesting residents such as Bonelli's Eagle and Black Wheatear can be found throughout the year, so a visit at any time can be rewarding. It is possible to visit the Serranía in a day while based in Gibraltar.

There are three main sites - Grazalema, Sierra de las Nieves and Sierra Blanca which between them hold all the breeding species of the Serranía.

Apart from these, the following two itineraries are complementary. They follow mountain roads and can be taken as alternatives by those with only a little time available but who want to sample the mountain habitats and typical birds of the area.

Itinerary 1: From Ronda travel north-west along the C 339 and turn left to Montejaque after 14km. Follow the MA 501 south-west past Benaoján and Cortes de la Frontera and join the CA 504 to Ubrique. This is a route with impressive mountain scenery and good opportunities to see mountain birds.

Itinerary 2: From Ronda travel south-east along the C 339 for 11.5km. Turn right towards Cartajima. The narrow, winding road passes through valleys wooded with oaks and chestnuts and then climbs past bare mountain tops, through the villages of Cartajima, Júzcar, Faraján and Alpandeire before joining the C 341. Turn left at the C 341 and travel south-west towards Algatocín. At Algatocín, turn left towards Estepona. The mountain road (MA 536) drops down to the valley of the Rio Genal, through oak and chestnut woodland, and then climbs through Maritime Pines to the peaks of the Sierra Bermeja. There is a small wood of Spanish Firs near the peak. From here, the road (MA 557) descends through coastal scrub towards Estepona on the coast.

Grazalema

The mountains around the town of Grazalema form the western boundary of the Serranía de Ronda. The highest peak in the Sierra del Pinar reaches 1654m. These limestone mountains are spectacular, with exposed scree, bare rock and spectacular cliffs which contrast with the wooded valleys of Holm Oak. The Grazalema sierras hold an interesting breeding bird community, typical of these mountains. The area described has been designated a Natural Park, covering 51,695 hectares. An important feature is the well preserved wood of endemic Spanish Firs.

Accommodation

A good place to stay is the Hostal de Grazalema. This is a small 2-star hotel situated on a hill overlooking the town. It is advisable to book in advance [telephones: (956) 141162, 141136 or 141187].

Key

Roads

Mountain Peaks

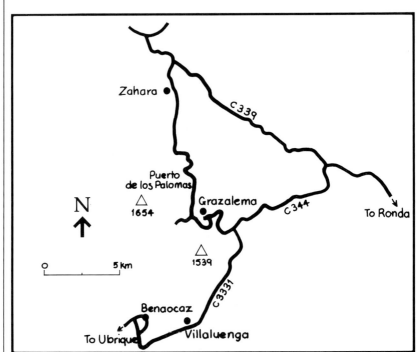

Strategy

The area is best in spring when the summer visitors add to the resident bird community. Other times can be good for finding some of the resident mountain species which are not found elsewhere in the area. The woodland site should be visited for species such as woodland raptors and Subalpine Warbler. The gorge is good for cliff nesting raptors, Choughs and specialities such as Black Wheatear and Rock Sparrow. The mountain pass has interesting breeding species including Rock Thrush along the highest points and Black Redstart which in southern Spain only breeds at high altitude.

Access to Grazalema is via mountain roads, whether from Gibraltar, Ronda or Jerez. The area can be visited in a day from Gibraltar but, to do these mountains justice, at least one night should be spent in Grazalema. Three sites within the mountains, close to Grazalema, are described - an oak wood in the valley, a rocky gorge and a mountain pass.

Woodland: Take the C 344 eastwards from Grazalema towards the C 339 which goes to Ronda. After 3.5km from Grazalema take the left turning to Ronda at the Puerto de los Alamillos. The road runs through oak woodland for about 11km until the junction with the C 339. Stop along this road at intervals to observe woodland species. Some areas of woodland are fenced but can be watched from the road, other areas are open and can be explored on foot.

Rocky gorge: Return to the Puerto de los Alamillos and turn left towards Villaluenga and Ubrique. The road (C 3331) runs for 15km to the town of Benaocaz and beyond that to Ubrique. The stretch to Benaocaz goes along the base of the impressive Sierras del Endrinal and del Caillo. Past the town of Villaluenga del Rosario, the road runs for 3km along the base of a steep gorge known as La Manga de Villaluenga. Stop at intervals along this road to check the mountains and cliffs as well as the valley which has a mixture of pasture and open woodland. It is possible to walk along paths into the mountains at various points.

Mountain pass: From Grazalema take the road west towards the mountains for 2km. At the road junction, take the right turning to Zahara de la Montaña (CA 531). This is a 20km stretch of narrow, winding, mountain road with spectacular scenery. The area is mostly upland scrub and cliff. The road climbs to the Puerto de las Palomas (1331m) and then gradually descends towards Zahara. 2km from the beginning of the road, a lay-by on the left marks the start of a track to the Spanish Fir wood. Permission must be obtained to enter the wood which is interesting but does not have any unusual birds. For permission to enter the wood write to the Agencia del Medio Ambiente, Avda. Ana de Villa, 3-3° izda, Cádiz [tel: (956) 274629] or by phoning the park office [tel: (956) 123114]. Stop at suitable lay-bys along the road and look for mountain birds. There are some paths which can be followed but most areas are steep and are best watched from the roadside.

Birds   Interesting breeding mountain species are: Egyptian and Griffon Vultures, Short-toed Eagle, Goshawk, Golden, Booted and Bonelli's

Black Wheatear

Eagles, Peregrine, Rock Dove, Scops, Eagle and Tawny Owls, Bee-eater, Green Woodpecker, Woodlark, Thekla Lark, Crag Martin, Tawny Pipit, Black Redstart, Black-eared and Black Wheatears, Rock and Blue Rock Thrushes, Melodious, Dartford, Subalpine, Orphean, and Bonelli's Warblers, Firecrest, Crested Tit, Short-toed Treecreeper, Golden Oriole, Great Grey and Woodchat Shrikes, Chough, Raven, Spotless Starling, Rock Sparrow, Serin, Hawfinch and Cirl and Rock Buntings.

Sierra de las Nieves

Torrecilla (1919m) is the highest peak of the Serranía de Ronda. Around it are other high peaks, and all are mostly bare or with low scrub, resistant to the harsh winter climate. Definite belts of vegetation can be seen at different heights up the mountain sides.

Sierra de las Nieves, a Natural Park (covering 18,530 hectares) just north-west of Torrecilla, is covered with Holm Oak woodland at around 1000m which is mature and extensive. At 1200m it gives way to Maritime Pine which forms woods of mature trees mixed with younger stands with scrub vegetation in between, including Hawthorn. Higher up, at around 1300m, on the north-facing slopes, impressive woods of tall, old, endemic Spanish Firs dominate the landscape. Beyond these, above 1500m, the landscape is barren, dominated by scree and low bushes with a few, scattered, Yews and Firs.

Such a rich complex of vegetation gives rise to a high diversity of bird species within the short distance of a single mountainside. In spring, the Sierra is full of birds, including many summer visitors which come to breed and leave in late summer before the harsh winter weather sets in.

Sierra de las Nieves is an excellent introduction to the Serranía de Ronda and has all the species which breed in these mountains. In addition, it has impressive scenery and some of the best panoramic views in the area.

Key

Pines

Oaks

Roads

Tracks

Mountain Peaks

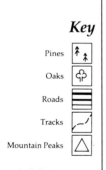

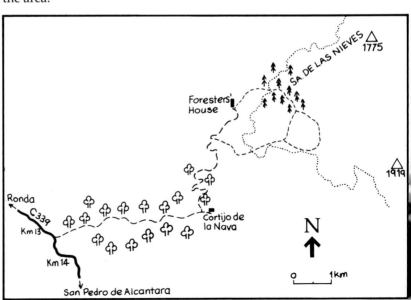

Accommodation

The site can be visited within a day from Gibraltar but there are hotels closer, at Ronda and San Pedro de Alcantara. Permission to camp in the sierra should be obtained from the Agencia del Medio Ambiente, Molina Larios, 13-2°, Málaga [tel: (952) 225800] although an enquiry at the forester's house may sometimes suffice.

Strategy

The Sierra de las Nieves is best in spring which arrives late in these mountains. The best time to visit is from late April onwards when the summer visitors will have arrived. A walk to the top of the mountain is necessary to see the complete range of species some of which, such as Rock Thrush, can only be found higher up.

From the C 339 between Ronda and San Pedro de Alcantara (on the Costa del Sol) take a track on the eastern side of the road between Km13 and Km14. The track, which in dry weather is firm and can be driven by car (4-wheel preferable but not essential), crosses the oak wood for about 6km. It is possible to stop anywhere along the track and wander into the woods in any direction with the exception of a small area which is fenced. To get to the higher ground, bear left where the track forks 2.2km from the main road. There is a white gate which is usually open at this point. At 3km from the main road there is a closed gate - open it and pass through making sure to close it afterwards. The track continues past a farm on the right at 4.5km and starts to ascend at this point, crossing a cattle grid at 5.5km. It continues through scrub and maritime pine until a fork is reached at 9.2km. Take the left fork down towards the foresters' house which is in a small valley and park by the house or in the nearby camping area.

A recommended walk from here is back to the fork at 9.2km, taking the other track which ascends through scrub and pine, eventually reaching Spanish Fir woodland. From here it is possible to follow smaller tracks through the fir wood which gradually descend towards the starting point. Alternatively, follow the main track which leads to the peak (4km from where the vehicle was left) from where it is possible to join the track through the fir wood back down or return along the main track. Those wanting to go on a longer hike can go cross country towards Torrecilla which lies to the south-east but a map and compass are essential.

Most of the birds of the sierra can be seen by following the recommended tracks and walking along the oak woods. The walk up the mountain side takes two to three hours depending on which route is chosen and the speed of walking. A whole day should be allowed to explore the sierra properly.

Birds

Interesting breeding species are: Egyptian and Griffon Vultures, Short-toed Eagle, Goshawk, Golden, Booted and Bonelli's Eagles, Peregrine, Eagle Owl, Quail, Alpine and White-rumped Swifts, Hoopoe, Green Woodpecker, Woodlark, Thekla Lark, Skylark (rare breeding species in southern Spain confined to mountain tops), Crag Martin, Tawny Pipit, Black Redstart, Redstart, Wheatear (rare breeder), Black-eared and Black Wheatears, Rock and Blue Rock Thrushes, Melodious, Dartford, Subalpine, Orphean and Bonelli's Warblers,

Firecrest, Crested and Coal Tits, Short-toed Treecreeper, Great Grey and Woodchat Shrikes, Chough, Raven, Rock Sparrow (frequently seen on a small cliff on the right at the lower end of the fir wood), Crossbill and Rock Bunting.

Other Wildlife   Spanish Ibex, Fire Salamander.

Sierra Blanca

At the south-eastern end of the Serranía de Ronda lies a coastal sierra known as the Sierra Blanca. It is a limestone formation with a maximum altitude of 1215m. It is close to the Mediterranean coast and there is a vegetation sequence from Mediterranean coastal scrub (garrigue) through Maritime Pine woodland (partly planted but native to the area) to upland scrub. There are areas of exposed rocks, especially near the peaks and a few interesting cliffs. Within the sierra there is a wide valley surrounded by mountains. Much of this is wooded, with stands of Sweet Chestnut and smaller ones of Eucalyptus and larger woods of Maritime Pine with scattered Holm Oaks. There is an olive grove in the centre of the valley.

Accommodation   The Refugio de Juanar is an ideal place to spend two or more days in the Sierra Blanca. It is also a good base for exploring other sierras in the area and some of the mountain towns such as Coín and Tolox. It is a 3-star hostel with limited capacity and bookings should be made in advance [telephones: (952) 881000 or 881001]. The nearest camping sites are on the coast.

Strategy   The sierra can be visited for the day from Gibraltar or one of the nearby coastal towns. However, if the mountains are to be explored then two days should be spent here. Spring is the best time to visit these mountains as the breeding species are of most interest. In winter, there are fewer birds although Ring Ouzel and Alpine Accentor can be seen.

Key

Refuge/Hostel   ■

Roads

Tracks

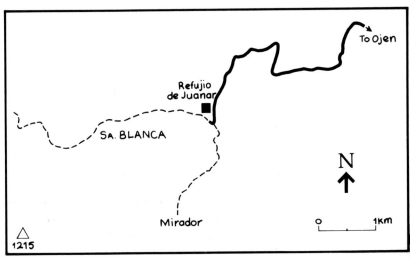

Some migrants occur on passage but the sierras are well away from the main flylines close to the Strait. In autumn in particular, some of the raptor streams which follow the coast towards Gibraltar can be seen from here.

The best approach is from the Costa del Sol. From Marbella, take the C 337 inland towards the mountain town of Ojén. The road climbs for 9km through a mixture of coastal scrub, Maritime Pine, Carob and exotic woodland. Past Ojén the road climbs steeply towards the Puerto de Ojén (580m). Continue along this road past a quarry on the left and take a left turning marked 'Refugio de Juanar'. This mountain road climbs through pine woods and scrub and reaches the Refugio, a small hostel. Park here and walk south along the track marked 'El Mirador'. This track takes you through pine woodland, then olive groves and, finally, coastal scrub. The view from the end of the track is spectacular. Return along the track and half way, on the edge of the olive groves, there are the ruins of an old farm. From here there is a path which leads north-west towards other pine woods and, eventually, leads to the main track near the refugio. There are also other paths leading off from the main track which can be explored.

An alternative walk, which is harder, goes towards one of the peaks. The mountain path starts just north-west of the refugio, behind the tennis court. It winds up the side of the mountain through upland scrub which holds Dartford Warblers. After about 1.5km the path changes direction northwards and the peak of Torrecilla becomes visible. The path continues towards the Serranía de Ronda and days could be spent wandering through these mountains. Unless specifically interested in continuing, this is a good point at which to turn back.

Birds   The sierra is best visited in spring for the breeding species which include Goshawk, Golden, Booted and Bonelli's Eagles, Peregrine, Alpine Swift, Green Woodpecker, Woodlark, Crag Martin, Red-rumped Swallow (in lower hills before Ojén), Blue Rock Thrush, Dartford, Subalpine, Orphean and Bonelli's Warblers, Firecrest, Crested Tit, Short-toed Treecreeper, Chough, Raven, Crossbill, Hawfinch and Rock Bunting.

Other Wildlife   Spanish Ibex.

Cabo de Gata, Almería

Almería is the south-eastern province of Andalucía. It is the most arid province and many areas are steppe- and desert-like. The proximity to similar areas of North Africa makes Almería an attractive area of southern Spain to visit and species not found elsewhere in Andalucía, or indeed Spain, can be seen. Cabo de Gata - Níjar is a Natural Park, covering 26,000 hectares.

Much of the province is treeless and covered by low scrub (garigue), steppe or desert. At its south-eastern corner lies the Cabo de Gata

Natural Park which has a combination of coastal, garigue-covered sierra with a spectacular rocky shore, steppe and a large salt pan complex. This variety of habitats makes the Cabo de Gata the most attractive area in the province with very interesting species including Dupont's Lark and Trumpeter Finch. Other steppe areas in the province are of interest as they contain the typical species of the province.

Almería is isolated from the rest of the areas included in this guide and an overnight stay should be contemplated if the area is to be included in an itinerary of bird-watching sites in southern Spain.

Accommodation   There is a good range of hotels in the area, which is being developed as a coastal resort. Try hotels in Carboneras (1 and 2-star), Níjar (1-star) and Mojacar (1-star to 4-star). The 4-star Parador Nacional at Mojacar is recommended. There are camping sites in Almería, San José (open June-September) and Mojacar.

Strategy   Almería is 330km from Gibraltar and cannot be visited on a day trip. Those attempting all the sites mentioned here should allow at least two full days. Those with less time should concentrate on Cabo de Gata and Campo de Níjar.

The area is best visited in spring (April/June) for the interesting breeding species and passage waders (on the Gata salt pans). Most bird-watching should be done in the morning and evening as the middle of the day, even in spring, can be very hot and birds become inactive. Early morning and late evening are the best times for Little Bustard, Stone-curlew and Black-bellied Sandgrouse.

From Almería, Cabo de Gata can be reached by following the N 332 east/north-east towards Vera and Murcia. After 14km, take a right turning to Cabo de Gata. This country road follows the coastal steppe areas which are dominated by low vegetation with succulents. Part of the area has signs indicating that it is a protected area and is excellent for steppe birds. The road crosses the Rambla de Morales (ramblas are seasonal river beds which usually attract many birds). Beyond the Rambla there is a fork in the road. The left turning can be taken towards the northern part of the Sierra de Cabo de Gata and the picturesque fishing village of San José.

If on a short visit, continue instead to Cabo de Gata, passing the village and then going south-east along the beach road towards the rocky cape which is visible in the distance. On the left, salt works will become apparent and the large, lagoon-like, salt pans. A stop here is recommended and it is possible to approach the pans along a number of tracks. There are usually large numbers of Greater Flamingoes here as well as Lesser Short-toed Larks and Spectacled Warblers. From here, the road leads to the sierra and the spectacular coast around the cape. Behind the cape, the steep hillsides are covered in low coastal scrub.

Other sites near Almería which are worth visiting include the Campo de Níjar area, the Sierra de Cabrera and the Desierto de Tabernas. The former is a vast area of steppe which has most of the typical steppe birds including Dupont's Lark and Trumpeter Finch. Campo de Níjar is the large area of steppe on the left of the main N 332 between the turn-off to

Cabo de Gata and the turn-off to Carboneras, 31km to the north-east. There are several roads which lead off the main road through the area and to the Sierra de Alhamilla behind the Campo de Níjar. The country road to Las Cuevas de los Medinas, west of the Cabo de Gata junction, is interesting. Many Rollers breed in this area.

The next area to visit is the spectacular Sierra de Cabrera. Take the right turning to Carboneras off the N 332 (AL 101) which reaches this coastal town after 19km. The road crosses another large area of steppe which can produce the typical birds of the habitat. From Carboneras, follow the narrow road which goes north, first along the shore, then through steep mountain valleys, for 22km to the pretty town of Mojacar.

From Mojacar follow the AL 151 west (which becomes the AL 150) for 14km to join the N 340. Follow the N340 south-west towards Almería for 48km. The road crosses the unique Desierto de Tabernas. There are tracks and country roads leading off from the main road into the desert and adjacent steppe areas which are worth a look. Trumpeter Finches breed in the desert. The right turning at Los Yesos (C 3325) after 39km from the AL 150/N 340 junction, and the right turning towards Castro de Filabres, a further 9km west, are worth exploring and lead to the interesting Sierra de los Filabres. These areas are good for breeding Rollers and Rock Sparrows. From Tabernas it is possible to return to Almería, having completed the circuit.

If travelling to and from Almería from Málaga along the N 340, the small lagoon complex known as the Albufera de Adra, 47km west of Almería, should be visited. This lagoon, which is visible from the main road, attracts passage and breeding waterbirds and has been recently colonised by White-headed Ducks.

Bee-eater

Birds   In spring, interesting breeding and passage species in the Cabo de
Gata/Níjar area are: Night and Purple Herons, Spoonbill, Greater
Flamingo, Garganey, Marbled Duck, Honey Buzzard, Short-toed Eagle,
Marsh and Montagu's Harriers, Bonelli's Eagle, Osprey, Peregrine
(Sierra de Cabrera), Quail, Water Rail, Spotted and Baillon's Crakes,
Little Bustard, Black-winged Stilt, Avocet, Stone-curlew, Little Ringed
and Kentish Plovers, Knot, Sanderling, Little and Temminck's Stints,
Ruff, Black-tailed and Bar-tailed Godwits, Spotted Redshank,
Greenshank, Great Skua, Mediterranean and Audouin's Gulls, Gull-
billed, Whiskered and Black Terns, Black-bellied Sandgrouse, Eagle and
Short-eared Owls, Red-necked Nightjar, Bee-eater, Roller, Hoopoe,
Wryneck, Dupont's, Calandra, Short-toed, Lesser Short-toed and Thekla
Larks, Woodlark, Crag Martin, Rufous Bush Robin, Black-eared and
Black Wheatears, Blue Rock Thrush, Cetti's, Great Reed, Dartford and
Spectacled Warblers, Great Grey and Woodchat Shrikes, Rock Sparrow,
Trumpeter Finch and Rock Bunting.

Other Wildlife   Natterjack Toad, Montpellier Snake, Ladder Snake, Southern Smooth
Snake, Horshoe Whip Snake, Lataste's Viper, Ocellated Lizard, Large
Psammodromus, Spanish Psammodromus, Spiny-footed Lizard,
Moorish Gecko, Turkish Gecko, Amphisbaenian, Iberian Wall Lizard,
Western Hedgehog, Algerian Hedgehog, Large Mouse-eared Bat,
Schreiber's Bat, Mediterranean Pine Vole, Weasel, Red Fox, Genet.

Key

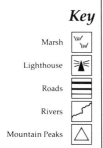

Marsh

Lighthouse

Roads

Rivers

Mountain Peaks

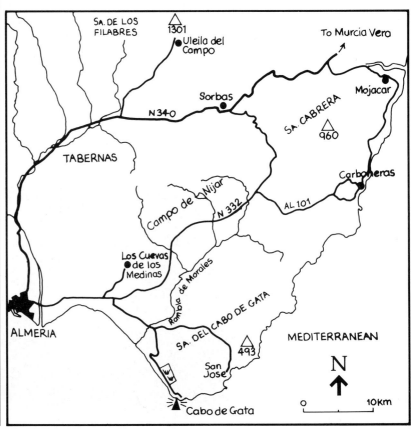

BRIEF NOTES ON ADDITIONAL SITES

The following notes provide supplementary information on a number of sites not included in the main site descriptions. These have been selected because they may be important for a species or group of species which may be difficult to find in the main sites described, because they are alternative sites to those described in detail for species which are typical of southern Iberia, or because they are considered important bird sites and have been recognised as such by official bodies. These notes are only supplementary and are designed to give an indication of their importance and location for the birdwatcher who has additional time to spend in the area. A number of these sites are some distance away from the nucleus of main sites but all are within the boundaries of the region of Andalucía.

Province of Cádiz

Cádiz Bay The large, shallow Bay of Cádiz and its associated salt marsh and salt pan complex are an important area for breeding, migratory and wintering waders and seabirds. The area is accessible from various points along the N IV between El Puerto de Santa María and Chiclana de la Frontera with the main area lying within a triangle formed by these two cities and Cádiz. 10,000 hectares have been declared a Natural Park.

Breeding birds include the largest concentration of Little Terns in Spain, as well as Kentish Plover, Black-winged Stilt and Avocet. Many waterbirds and waders occur on passage and in winter.

Lesser Crested Tern

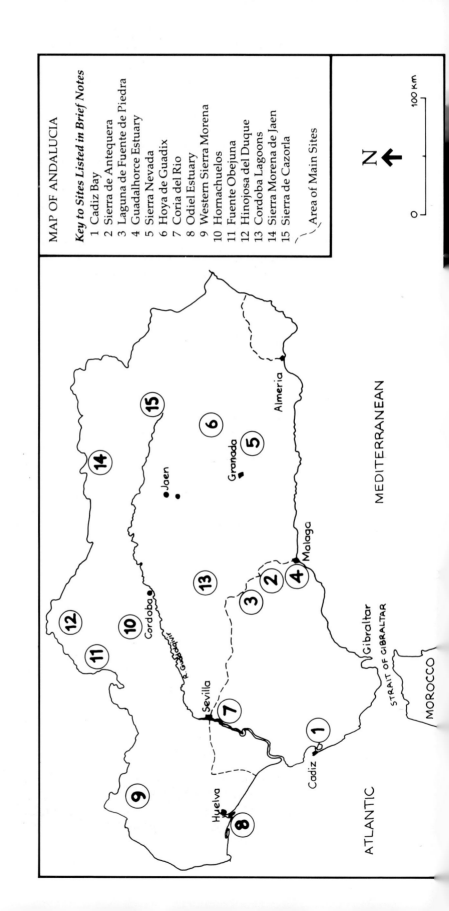

MAP OF ANDALUCIA

Key to Sites Listed in Brief Notes

1 Cadiz Bay
2 Sierra de Antequera
3 Laguna de Fuente de Piedra
4 Guadalhorce Estuary
5 Sierra Nevada
6 Hoya de Guadix
7 Coria del Rio
8 Odiel Estuary
9 Western Sierra Morena
10 Hornachuelos
11 Fuente Obejuna
12 Hinojosa del Duque
13 Cordoba Lagoons
14 Sierra Morena de Jaen
15 Sierra de Cazorla

‒‒‒ Area of Main Sites

Province of Málaga

These sites can easily be visited if based in Málaga where there is a wide range of accommodation. Antequera has 1-star to 3-star accommodation including a Parador Nacional.

Sierra de Antequera

These mountains, which lie north and north-west of Málaga, consist of open limestone areas and woods of Holm Oak and Stone and Aleppo Pine. The area has the status of Natural Park.

From Málaga, the MA 402 leads north-west to Alora from where a country road leads north-west to the impressive gorge known as the Garganta del Chorro or Desfiladero de los Gaitanes.

From Alora the C 337 goes north towards Antequera and the C 3310 turns right off this to the Torcal de Antequera with its impressive rock formations. From here continue south along the MA 424 to Málaga.

Interesting breeding species include Griffon Vulture, Golden and Bonelli's Eagles, Peregrine, Rock Dove, Alpine Swift, Crag Martin, Black Wheatear, Blue Rock Thrush, Golden Oriole, Chough and Raven. Chamaeleons are also present.

Laguna de Fuente de Piedra

A large (1,364 hectares), shallow lagoon and a number of smaller lagoons surrounded by agricultural land. The area has the status of Natural Reserve.

To reach these lagoons follow the N 334 north-west from Antequera towards Sevilla. The village of Fuente de Piedra is reached after 21 km. Drive through the village, cross a railway track and the lagoon is visible on the left.

Fuente de Piedra has Europe's second largest breeding colony of Greater Flamingoes (up to 12,000 pairs). Other breeding species on or around the lagoon include Montagu's Harrier, Black-winged Stilt, Avocet, Stone-curlew, Slender-billed Gull and Gull-billed Tern. Cranes winter in this area.

Guadalhorce Estuary

An isolated estuary along the Mediterranean Costa del Sol, west of Málaga. The estuary, in spite of pollution and urbanisation, has typical riverside vegetation and small ponds which attract migrants, especially waders and aquatic species which find few such areas of habitat in the area.

From Torremolinos, travel east along the main road towards the river and take one of a number of turn-offs to the right which lead to the estuary or the adjacent beach.

Province of Granada

Sierra Nevada

This is the second highest mountain range in Europe with a maximum altitude of 3482m at Mulhacén. Much of the range, especially near the peaks, is barren but there are important areas of alpine vegetation as well as scrub and some pine and broadleaved woodland. The Sierra is a Natural Park covering 169,239 hectares.

From Granada, take the GR 420 south-east towards Sierra Nevada. The climb to the Pico Veleta (3398 m), close to Mulhacén is 46km. The barren areas around these peaks are the most interesting. From Veleta it is possible to follow the GR 411 down towards the Alpujarras village of Capileira and the main N 323 via Lanjaron but this road is in poor ´ condition.

Granada has a good range of hotels and there are several hotels (2-star to 4-star) on the peaks of Sierra Nevada including a Parador Nacional. On the south slopes of the Sierra there are hotels (usually 1- and 2-star) in Lanjaron, Orjiva Pampaneira, Trevelez and Ugijar.

The two main breeding species which may be hard to find elsewhere are Alpine Accentor and Ortolan Bunting, which also occur lower down on passage or in winter. Other breeding species of the Sierra include Golden and Bonelli's Eagles, Peregrine, Eagle Owl, Crag Martin, Subalpine Warbler, Black Wheatear, Blue Rock Thrush, Rock Thrush, Rock Bunting and Chough.

Hoya de Guadix

A dry depression north-east of Granada dominated by semi-arid steppe and low scrub and savanna-like areas with Holm Oaks and some areas devoted to agriculture.

From Granada take the N 342 north-east to the town of Guadix (58km). From Guadix, continue north-east along the N 342 towards Baza. This road passes the Hoya on the left and the Sierra de Baza on the right. 18km from Guadix there is a country road on the left which runs through the Hoya for about 23km, past the village of Gorafe, to Baños de Alicún de las Torres and follows the Rio Gor. From Baños de Alicún continue south, back to Guadix via Fonelas and Benalúa de Guadix (31km) past open countryside. There are other tracks in the area which enter various parts of the Hoya. There is also a road which leads south-east from the main road for 6.5km to Gor, which may be worth visiting. There is 1- and 2-star accommodation in Guadix.

The Hoya is a rich area for steppe specialities as well as other species, including: Egyptian Vulture, Goshawk, Montagu's Harrier, Peregrine, Hobby, Quail, Little Bustard, Stone-curlew, Black-bellied Sandgrouse, Rock Dove, Eagle and Long-eared Owls, Green Woodpecker, Great Spotted Cuckoo, Roller, Hoopoe, Golden Oriole, Great Grey Shrike, Spectacled Warbler, Black Wheatear, Raven, Carrion Crow, Chough and Tree and Rock Sparrows. Beyond Baza is the Hoya de Baza, another arid depression with good numbers of Lesser Short-toed Larks and a few Dupont's Larks.

Province of Sevilla

Coria del Rio

The Guadalquivir passes south-west past Sevilla towards Coria del Rio at the northern edge of the Marismas del Guadalquivir. The area is good for marsh birds typical of the Doñana area and there is a mixed heronry on a river island by Coria del Rio.

From Sevilla take the SE 660 to Coria del Rio. A back road follows the river to La Puebla del Rio and this road can be followed further for some distance along the border of the marismas. Among the interesting breeding species are Night and Squacco Herons.

Province of Huelva

Odiel Estuary

The River Odiel opens into the Atlantic west of Huelva. The estuary which has saltmarsh, sandy spits and beaches, attracts passage and breeding waders and waterbirds. The site is a Biosphere Reserve, covering 7,185 hectares.

From Huelva take the H 414 west across the river and take the first left turning down a road which follows the west bank of the Odiel down to the sea. There are many interesting places to stop from which to observe the marsh and estuary. There are hotels in Huelva but the visitor staying around the Coto Doñana can visit Odiel from there.

Interesting breeding species include a colony of ground nesting Spoonbills, estimated at 300 pairs, Purple Heron, Black-winged Stilt and Kentish Plover. Important passage and wintering species include: Black-necked Grebe, White Stork, Greater Flamingo, Shelduck, Pintail, Garganey, Red-crested Pochard, Red-breasted Merganser, Red Kite, Marsh and Montagu's Harriers, Osprey, Peregrine, Avocet, Stone-curlew, Little Ringed Plover, Knot, Sanderling, Little Stint, Curlew Sandpiper, Black-tailed and Bar-tailed Godwits, Whimbrel, Spotted Redshank, Greenshank, Green and Wood Sandpipers, Hoopoe and Great Grey Shrike. Chameleon occur in the area.

Western Sierra Morena

The western limits of the Sierra Morena in Huelva near the border with Portugal are formed by the Sierras Pelada and de Aracena together with the Picos de Aroche. The sierra is dominated by open oak woodland (dehesas) and chestnut woodland. It is a Natural Park covering 184,000 hectares.

From Huelva follow the N 435 north, past Valverde del Camino. 67km north of Valverde there is an intersection. The N 433 runs west towards Portugal and east towards Aracena. The road west is poor and

Black Vulture

leads north towards the Pico de Aroche, 24.5km after the intersection. There are other mountain roads around Aroche and Aracena which can be explored with time. There is 1- and 2-star accommodation in Aracena.

The main interest of these sierras is that they hold the largest breeding population of Black Vultures in Andalucía (estimated at 30 pairs) and a pair of Spanish Imperial Eagles. Mammals include Wolf, Lynx, Otter, Red Deer, Egyptian Mongoose and Wild Boar.

Province of Córdoba

Hornachuelos

The Sierra Morena in Córdoba is a low mountain range not exceeding 680m. The area includes typical Mediterranean oak woodland, scrub and pasture. It is a Natural Park covering 67,202 hectares.

From Córdoba travel west along the C 431 for 45km. Turn right on the CO 142 for 8km to Hornachuelos. Areas of interest are around the Los Angeles monastery, near the Bembézar Dam, around Las Aljabaras and the woodland bordering the Guadalora River. There is 1- and 3-star accommodation in Palma del Rio.

The area has Andalucía's second largest breeding population of Black Vulture (at least 20 pairs) as well as Griffon Vulture and Golden and Bonelli's Eagles. Spanish Imperial Eagle and Black Stork have been reported. Mammals include Lynx, Egyptian Mongoose, Otter, Red Deer, Wild Boar and a few Wolves.

Fuente Obejuna

This is a low, undulating, area with grassland and open oak woodland.

From Córdoba follow the N 432 north-west towards Peñarroya Pueblonuevo and continue to Fuente Obejuna, a total of 96km. From Fuente Obejuna a country road leads north-east for 15km to the village of La Granjuela. From here a country road leads north-west for 9km to the village of Los Blázquez from where another road leads south back to Fuente Obejuna. 1-star accommodation is available in Peñarroya Pueblonuevo.

The area is an important passage and wintering site for Crane; about two dozen Great Bustards and other steppe species breed.

Hinojosa del Duque

Close to Fuenta Obejuna is a similar area of steppe. From Peñarroya Pueblonuevo take the C 421 north-east for 13km and turn left along the CO 440 for 18km towards Hinojosa del Duque. Follow an 18km country road east from Hinojosa to El Viso. There is 1-star accommodation at Hinojosa del Duque.

The area between Hinojosa and El Viso is a passage and wintering area for Crane and is also attractive to steppe breeding birds. Although Great Bustards are scarce there are many Little Bustards in this area.

Córdoba Lagoons

In the south of the province of Córdoba there is a complex of permanent and semi-permanent shallow lagoons which are rich in breeding and wintering waterbirds. There are six main Lagunas (lagoons) within a small area - Zoñar, Amarga, del Rincón, del Salobral, Tiscar and de los Jarales. They are all Natural Reserves.

From Córdoba take the N IV south for 15km and then take the left fork (N 331) south-east to Aguilar (37km). From Aguilar take the C 329 south-west which passes the Laguna de Zoñar on the right after approximately 4km. Continue towards Puente Genil and take a right hand turning along a country road just before the town. Follow this road for 7km, passing the village of Puerto Alegre before reaching the Laguna de Tiscar. From Aguilar take the CO 760 south-east to Moriles. The Laguna del Rincón is on the left after 7km. These three lagoons are representative of the Córdoba lagoons. Lagunas Amarga and de los Jarales are south-west of Lucena which is 20km south-east of Aguilar and Laguna del Salobral is south-east of Baena which lies north-east of Aguilar. Aguilar has 1-star accommodation and there is 2-star accommodation in Lucena, Puente Genil and Baena. From the south, these lagoons can be approached from Antequera. Those wanting further information should write to the local society - Amigos de la Malvasía, Apartado de Correos 3.059, Córdoba.

Interesting breeding species are Purple Heron, White-headed Duck, Purple Gallinule, Black-winged Stilt and Avocet. Wintering and passage species include Greater Flamingo, Red-crested Pochard and Marsh Harrier.

Province of Jaén

Sierra Morena de Jaén

At the northern edge of Jaén, on the limits of Andalucía, a vast area of Mediterranean woodland and scrub stretches from Andújar in the south-west to Despeñaperros, Aldeaquemada and the Embalse de Dañador in the north-east. There are several options for exploring this very large mountain range which is a Natural Park (Andújar - 60,800 hectares; Despeñaperros - 8,000 hectares).

From Jaén, take the N 323 north to Bailén. From here travel west along the N IV for 27km to Andújar. Then take the scenic mountain road J 501 which follows the western part of this sierra for 87km to the town of La Carolina. Two side roads, one after 15km to Santuario de la Virgen de la Cabeza and another a few kilometres further on to the Embalse del Jándula, are worth exploring. From La Carolina continue north along the N IV towards Despeñaperros. Before Despeñaperros, take either the J 612 from Santa Elena or the J 611 from Las Correderas to Aldeaquemada and beyond towards Castellar de Santiago (49km from main road). There are tracks off the road worth exploring. Alternatively, from La Carolina take the C 3217 south-east for 24km to Arquillos and then follow the C 3210 for 33km to a left turn near Castellar de Santisteban. This road (J 613), which becomes the J 614/621/620 leads to the Embalse de Dañador area and there are tracks off the road in several places. 1- to 3-star accommodation is available in Andújar.

Interesting breeding species include Black Stork, Black Vulture, Spanish Imperial, Golden and Bonelli's Eagles and Eagle Owl. Mammals include Wolf, Lynx, Red, Roe and Fallow Deer, Mouflon, Wild Boar and Egyptian Mongoose.

Sierras de Cazorla, Segura and Las Villas

A mountain range which is well wooded with oak and pine as well as scrub. The area is a Natural Park covering 214,000 hectares.

Take the N 321 from Jaén north-east to Ubeda. Continue along the N322 and turn right onto the J 314 past Torreperogil, 11km outside Ubeda. Follow this road south-east for 33km to Peal de Becerro and take the left road (C 328) to Cazorla (14km). Follow the road past Cazorla which leads into the park. There is a Parador Nacional (3-star) within the sierra and 1-star establishments in Cazorla and La Iruela.

Interesting breeding species include Griffon Vulture, Golden and Booted Eagles, Peregrine, Eagle Owl and Chough. Mammals include Otter, Red Deer, Mouflon, Ibex and Wild Boar.

SELECTIVE BIRD LIST

The following list provides additional information about a selection of local or uncommon species which can be difficult to find.

Storm Petrel. Present in the Strait between April and September but usually well offshore and not visible from the coast. A trip on the ferry from Algeciras to Tangier usually produces this species along the western part of the Strait.

Leach's Storm Petrel. A winter visitor to Atlantic areas adjacent to south-western Spain although it rarely comes inshore except after severe westerly or south-westerly gales when it may be numerous.

Little Bittern. A summer visitor which may be hard to observe because of its shy habits. It breeds in reed beds on the fringes of most lagoons (Medina, Salada, Pachecas, etc. – (p42) and on similar vegetation along drainage canals in La Janda (p36) and the Marismas (p47).

Night Heron. Occurs in aquatic habitats on passage and is to be found breeding in the Marismas and on the marshy ground around the Bornos Reservoir (Cádiz).

Squacco Heron. A very localised summer visitor which breeds within Coto Doñana (p47) and can be observed in peripheral areas around the park. Also breeds in a heronry on the river island at Coria del Rio (p64), south-west of Sevilla. Occurs in aquatic habitats on passage.

Purple Heron. Breeds in dense reed beds in the Marismas (p47), including along drainage canals, and on the Odiel Estuary (p65). A few pairs also nest in the denser reed beds along the fringes of some lagoons. It is regularly observed in other aquatic habitats on passage.

Black Stork. The southernmost nesting sites are in the Sierra Morena (p67), along the northern fringe of Andalucía. It is, nevertheless, frequently seen on passage over the Strait, sometimes in flocks. Main passage periods are late February-April and September-October.

Glossy Ibis. Irregular, although it has bred in the Marismas (p47) and near Huelva. It is possible in aquatic habitats and has occurred recently on Tarifa Beach (p31).

Spoonbill. Breeds within Coto Doñana (p47) and on the Odiel Estuary (p65). It is regularly observed in various parts of the Marismas (p47) including the salt pans at Bonanza (p45) and is recorded on passage in other sites, e.g. Barbate Estuary (p34).

Marbled Duck. Breeds in the Marismas (p47) where it can usually be found. Occurs in the Cádiz lagoons (p42) and the Palmones Estuary (p24) irregularly at any time of year, sometimes in large flocks.

Ferruginous Duck. Very rare. Occasionally observed in the Cádiz lagoons (p42) and probably in other aquatic habitats.

White-headed Duck. Breeds in the Cádiz (p42) and Córdoba lagoons (p66) where it may be found in large numbers. Laguna de Medina and Laguna Salada in Cádiz are recommended sites.

Black-shouldered Kite. Very rare. Has been regularly observed in winter at Laguna de La Janda (p36) in recent years. The main breeding area in Spain is in Extremadura.

Red Kite. Breeds in pine woods around the Marismas (p47) and is more widespread in lowland areas in winter. La Janda (p36) is a regular wintering site. Although scarce, it can be observed on passage across the Strait.

Black Vulture. Breeds in the Sierra Morena (pp65 & 67) along the northern fringe of Andalucía. Wanders right down to the shores of the Strait in winter and sometimes crosses into Morocco. Winter sites in recent years have included La Janda (p36), Alcalá de los Gazules (p41) and the Tarifa rubbish tip.

Goshawk. Widespread as a breeding bird in broadleaved woods although difficult to see. Regular sites are Almoraima (p28), Sierra Bermeja (Cádiz), Sierra de las Nieves (p54) and Sierra Blanca (p56). Very rare on passage.

Spanish Imperial Eagle. Breeds within and around Coto Doñana (p47) and in the Sierra Morena (pp65 & 67). Seen elsewhere outside the breeding season. La Janda (p36) is a regular wintering site.

Golden Eagle. Breeds in high mountain areas. Regular sites include Grazalema (p51), Sierra de las Nieves (p54) and Sierra Blanca (p56). Spreads to lower ground in winter including La Janda (p34).

Bonelli's Eagle. Breeds in many cliffs on high ground and near the coast. Regular sites include Sierra Crestellina (Málaga), Peñon de Zaframagón (Cádiz) and the mountains around Grazalema and Ronda (p51).

Merlin. Localised in winter on low ground, e.g. Marismas (p47) and La Janda (p36). Scarce on passage across the Strait.

Hobby. Breeds in Pinar de Monte Algaida (p45) and occurs regularly on passage across the Strait.

Eleonora's Falcon. Scarce but regular in the Strait area between April and October, especially in August and September.

Lanner. Rare on passage across the Strait in May, August and September.

Baillon's Crake. Difficult to observe in reed beds around lagoons and along drainage canals in Marismas (p47) and La Janda (p36) in spring where it probably breeds.

Purple Gallinule. Resident in reed beds around lagoons and in Marismas (p47). Easily observed in Laguna de las Pachecas (p42) and at El Acebuche (p47). Occurs elsewhere outside the breeding season. The Laguna de Torreguadiaro (p24) has been a regular site in recent winters.

Crested Coot. A rare breeding species which can be observed in Laguna de Medina and/or Laguna Salada (p42).

Crane. Nowadays occurs on passage and in winter. Up to 500 have been counted in winter at La Janda (p36) with greater concentrations in autumn. Other winter sites include the areas around Hinojosa del Duque and Fuente Obejuna in the north of Córdoba Province (p66).

Great Bustard. Very scarce and localised in the La Janda area around Tahivilla. A few are said to breed in the grasslands around Fuente Obejuna in Córdoba (p66).

Purple Sandpiper. Very rare in winter. A regular wintering site is Punta Secreta/Carnero at the south-western corner of the Bay of Gibraltar (p21).

Slender-billed Gull. Very localised as a breeding species with regular sites in the salt pans at Bonanza (p45) and the Laguna de Fuente de Piedra (p63).

Gull-billed Tern. Very localised as a breeding species with regular sites in the salt pans at Bonanza (p45) and the Laguna de Fuente de Piedra (p63). Regular, though scarce, on passage off Europa Point, Gibraltar (p19), in July and August.

Caspian Tern. Regular in the salt pans at Bonanza (p45) and, in winter, around Barbate (p34).

Royal Tern. Regular in Tangier in August and September and occasionally wanders to the European shore of the Strait where it has been recorded as far east as Europa Point, Gibraltar (p19).

Lesser Crested Tern. Regular but scarce in the Strait in September and October, occasionally April, with recent sightings off Europa Point (p19), in the Bay of Gibraltar (p21) and at Tarifa Beach (p31).

Black-bellied Sandgrouse. Regular at several sites in steppe habitat in Almería, including Cabo de Gata, Campo de Níjar and Desierto de Tabernas (p57) and in the steppe areas around Guadix (p64), north-east of Granada .

Pin-tailed Sandgrouse. Regular in the Marismas (p47). The east bank o the Guadalquivir (p45) around Trebujena is a regular site.

Great Spotted Cuckoo. Breeds in pine woods around the Marismas (p47) including the Pinar de Monte Algaida (p45) and Coto del Rey (p47). Seen elsewhere on passage from January to March and in July and August.

Eagle Owl. Breeds at many cliff sites and sometimes in trees in the area Regular cliffs in the province of Cádiz include Peñon de Zaframagón, Laja de la Zarga (Sierra Plata) and Peñon de Lagarin.

Red-necked Nightjar. Breeds in open pine woods and other light woodland. Regular sites include Pinar del Rey (p29) and Pinar de Monte Algaida (p45).

White-rumped Swift. Localised as a breeding species between late Ma and October in scattered sites in the area including Sierra de la Plata (Cádiz), Almoraima (p28) and the Jara Valley (p33), occupying nests of Red-rumped Swallows. Also found in high mountain areas, on Sierra de las Nieves (p54) and around Grazalema (p51).

Roller. Breeds in steppe areas of Almería and Granada where it can be easily observed along roadsides. Closer to the Strait occurs in lowland areas (e.g.La Janda [p36]) on passage in April, May, August and September but does not breed.

Dupont's Lark. Scarce as a breeding species in steppe areas in Almería around Campo de Níjar and Cabo de Gata (p57).

Lesser Short-toed Lark. Localised but in places abundant. Regular site are the steppe and marshy areas around the Marismas (p47), including the east bank of the Guadalquivir around Trebujena and Bonanza (p45 Abundant in the steppe and beach areas around Cabo de Gata, Almería (p57).

Thekla Lark. An abundant breeding species in open, hilly countryside throughout the area.

Alpine Accentor. Breeds along the highest peaks of the Sierra Nevada, Granada (p63). In winter found on lower ground including the peaks o the Rock of Gibraltar (p15) where it has wintered for many years.

Rufous Bush Robin. Localised as a breeding species in light woodland and orchards. Traditional sites are in the open woodland along the roa from Tarifa to Vejer and on some orchards in the Hozgarganta valley (p24). Can be observed elsewhere (e.g.Gibraltar) on passage.

Bluethroat. Scarce on passage and in winter. Recent wintering sites are the Guadiaro and Palmones Estuaries (p24), in reed beds.

Black Wheatear. A localised breeding species which is common in limestone rocky outcrops and cliffs, right down to sea level. It is best looked for in the eastern Andalucían province.

Rock Thrush. Breeds in high mountains above 1300 metres. Regular sites are Sierra de las Nieves (p54), the road between Grazalema and Zahara de la Montaña (p51), open scree areas between Faraján and Ronda (p51) and on the Sierra Nevada (p63). Scarce but regular on passage on the Rock of Gibraltar (p15), in April and September.

Savi's Warbler. Very localised in reed beds. A regular site is the reed bed on the Arroyo de la Rocina near El Rocio (p47).

Olivaceous Warbler. A very localised breeding species, usually in light woodland. Recent breeding sites are on the Jara Valley (p33) and the Almoraima (p28). Regular but scarce on passage, in April, May and August on Gibraltar (p15).

Spectacled Warbler. A localised breeding species in low coastal scrub and in the drier parts of marshes. Regular sites are the coastal scrub north-west of Sabinillas, the low scrub around Bonanza and Trebujena (p45) and the coastal scrub around Cabo de Gata in Almería (p57). Occurs as a scarce but regular migrant on Gibraltar (p15) in April and September.

Orphean Warbler. Localised as a breeding species in light woodland and orchards. Recent breeding sites have included Almoraima (p28), Pinar del Rey (p29) and Sierra de las Nieves (p54). Regular but scarce on passage at Gibraltar (p15) in April, May, August and September.

Penduline Tit. A regular winter visitor. Wintering sites include the Guadiaro, Palmones and Guadarranque Estuaries (p24), Laguna de Medina (p45) and the aquatic vegetation on the fringes of the drainage canals at La Janda (p36).

Azure-winged Magpie. Localised but abundant in the north-west. Regular sites are Pinar de Monte Algaida (p45), Coto del Rey (p47) and most pine woods in and around the Coto Doñana (p47).

Crossbill. Local in pine and fir woods in the mountains. Regular breeding sites are Sierra de las Nieves (p54) and Sierra Blanca (p56).

Trumpeter Finch. A scarce breeding species confined to steppe areas of Almería (p57). Regular sites are Campo de Níjar, Cabo de Gata and Desierto de Tabernas.

FULL SPECIES LIST

The following list includes all of the bird species known by the author to have been recorded in Gibraltar and Andalucía. The symbols for status and abundance refer to each species range within the area.

Key to checklist

R	resident
S	summer visitor
W	winter visitor
M	passage migrant
V	vagrant
E	extinct
L	localised distribution
1	common
2	fairly common
3	occasional
4	uncommon
5	rare
?	status uncertain

4W ☐	Red-throated Diver (Gavia stellata)
5W ☐	Black-throated Diver (Gavia arctica)
5W ☐	Great Northern Diver (Gavia immer)
1R ☐	Little Grebe (Tachybaptus ruficollis)
1R ☐	Great Crested Grebe (Podiceps cristatus)
5W ☐	Slavonian Grebe (Podiceps auritus)
1R ☐	Black-necked Grebe (Podiceps nigricollis)
V ☐	Bulwer's Petrel (Bulweria bulwerii)
1SM ☐	Cory's Shearwater (Calonectris diomedea)
3M ☐	Great Shearwater (Puffinus gravis)
3M ☐	Sooty Shearwater (Puffinus griseus)
1MW ☐	Balearic Shearwater (Puffinus [puffinus] yelkouan)
V ☐	Little Shearwater (Puffinus assimilis)
5M ☐	Wilson's Petrel (Oceanites oceanicus)
2SM ☐	Storm Petrel (Hydrobates pelagicus)
3W ☐	Leach's Petrel (Oceanodroma leucorhoa)
1MW ☐	Gannet (Sula bassana)
V ☐	Cape Gannet (Sula capensis)
V ☐	Masked Booby (Sula dactylatra)
2W ☐	Cormorant (Phalacrocorax carbo)
5RL ☐	Shag (Phalacrocorax aristotelis)
?E ☐	Bittern (Botaurus stellaris)
2SM ☐	Little Bittern (Ixobrychus minutus)
2SML ☐	Night Heron (Nycticorax nycticorax)
4SML ☐	Squacco Heron (Ardeola ralloides)
1R ☐	Cattle Egret (Bubulcus ibis)
1RM ☐	Little Egret (Egretta garzetta)
1MW ☐	Grey Heron (Ardea cinerea)
2SM ☐	Purple Heron (Ardea purpurea)

2M5SL ☐	Black Stork (Ciconia nigra)
1SM ☐	White Stork (Ciconia ciconia)
5SM ☐	Glossy Ibis (Plegadis falcinellus)
2SML ☐	Spoonbill (Platalea leucorodia)
1SML ☐	Greater Flamingo (Phoenicopterus ruber)
V ☐	Whooper Swan (Cygnus cygnus)
5W ☐	Bean Goose (Anser fabalis)
V ☐	White-fronted Goose (Anser albifrons)
1WL ☐	Greylag Goose (Anser anser)
5M? ☐	Ruddy Shelduck (Tadorna ferruginea)
2WL ☐	Shelduck (Tadorna tadorna)
1WM ☐	Wigeon (Anas penelope)
1RMW ☐	Gadwall (Anas strepera)
1MW ☐	Teal (Anas crecca)
1RMW ☐	Mallard (Anas platyrhynchos)
2MW ☐	Pintail (Anas acuta)
2M ☐	Garganey (Anas querquedula)
1MW ☐	Shoveler (Anas clypeata)
4RML ☐	Marbled Duck (Marmaronetta angustirostris)
2RMW ☐	Red-crested Pochard (Netta rufina)
2R1MW ☐	Pochard (Aythya ferina)
V ☐	Ring-necked Duck (Aythya collaris)
5R? ☐	Ferruginous Duck (Aythya nyroca)
4MW ☐	Tufted Duck (Aythya fuligula)
V ☐	Scaup (Aythya marila)
V ☐	Eider (Somateria mollissima)
V ☐	Long-tailed Duck (Clangula hyemalis)
1MWL ☐	Common Scoter (Melanitta nigra)
V ☐	Velvet Scoter (Melanitta fusca)
V ☐	Goldeneye (Bucephala clangula)
V ☐	Smew (Mergus albellus)
2WL ☐	Red-breasted Merganser (Mergus serrator)
V ☐	Goosander (Mergus merganser)
V ☐	Ruddy Duck (Oxyura jamaicensis)
2RL ☐	White-headed Duck (Oxyura leucocephala)
1M ☐	Honey Buzzard (Pernis apivorus)
3WM ☐	Black-shouldered Kite (Elanus caeruleus)
1SM ☐	Black Kite (Milvus migrans)
2RL2MW ☐	Red Kite (Milvus milvus)
E ☐	Lammergeier ((Gypaetus barbatus)
2SM ☐	Egyptian Vulture (Neophron percnopterus)
1RM ☐	Griffon Vulture (Gyps fulvus)
4RLMW ☐	Black Vulture (Aegypius monachus)
1SM ☐	Short-toed Eagle (Circaetus gallicus)
2RMW ☐	Marsh Harrier (Circus aeruginosus)
2MW ☐	Hen Harrier (Circus cyaneus)
V ☐	Pallid Harrier (Circus macrourus)
1SM ☐	Montagu's Harrier (Circus pygargus)
2RL ☐	Goshawk (Accipiter gentilis)
2RL1MW ☐	Sparrowhawk (Accipiter nisus)

1RMW ☐	Buzzard (Buteo buteo)
V ☐	Long-legged Buzzard (Buteo rufinus)
V ☐	Rough-legged Buzzard (Buteo lagopus)
V ☐	Lesser Spotted Eagle (Aquila pomarina)
5M ☐	Spotted Eagle (Aquila clanga)
V ☐	Tawny Eagle (Aquila rapax)
4RLM ☐	Spanish Imperial Eagle (Aquila [heliaca] adalberti)
2RL ☐	Golden Eagle (Aquila chrysaetos)
1SM4W ☐	Booted Eagle (Hieraaetus pennatus)
2RL ☐	Bonelli's Eagle (Hieraaetus fasciatus)
2MW ☐	Osprey (Pandion haliaetus)
1SM ☐	Lesser Kestrel (Falco naumanni)
1RMW ☐	Kestrel (Falco tinnunculus)
V ☐	Red-footed Falcon (Falco vespertinus)
4W ☐	Merlin (Falco columbarius)
4SL2M ☐	Hobby (Falco subbuteo)
3M ☐	Eleonora's Falcon (Falco eleonorae)
5M ☐	Lanner (Falco biarmicus)
1RL ☐	Peregrine (Falco peregrinus)
1R ☐	Red-legged Partridge (Alectoris rufa)
1RL ☐	Barbary Partridge (Alectoris barbara)
2SLM ☐	Quail (Coturnix coturnix)
1RL ☐	Pheasant (Phasianus colchicus)
E? ☐	Andalusian Hemipode (Turnix sylvatica)
2WML ☐	Water Rail (Rallus aquaticus)
5M ☐	Spotted Crake (Porzana porzana)
5M ☐	Little Crake (Porzana parva)
4SM ☐	Baillon's Crake (Porzana pusilla)
4M ☐	Corncrake (Crex crex)
1R ☐	Moorhen (Gallinula chloropus)
2RL ☐	Purple Gallinule (Porphyrio porphyrio)
1R ☐	Coot (Fulica atra)
5RL ☐	Crested Coot (Fulica cristata)
2MWL ☐	Crane (Grus grus)
V ☐	Demoiselle Crane (Anthropoides virgo)
1RL ☐	Little Bustard (Tetrax tetrax)
5RL ☐	Great Bustard (Otis tarda)
2MW ☐	Oystercatcher (Haematopus ostralegus)
1SM ☐	Black-winged Stilt (Himantopus himantopus)
1RML ☐	Avocet (Recurvirostra avocetta)
2RML ☐	Stone-curlew (Burhinus oedicnemus)
V ☐	Cream-coloured Courser (Cursorius cursor)
1SL ☐	Collared Pratincole (Glareola pratincola)
V ☐	Black-winged Pratincole (Glareola nordmanni)
2SLM ☐	Little Ringed Plover (Charadrius dubius)
1MW ☐	Ringed Plover (Charadrius hiaticula)
1R ☐	Kentish Plover (Charadrius alexandrinus)
5M ☐	Dotterel (Charadrius morinellus)
V ☐	Lesser Golden Plover (Pluvialis dominica)
1WL ☐	Golden Plover (Pluvialis apricaria)

1MW ☐	Grey Plover (Pluvialis squatarola)
V ☐	Spur-winged Plover (Holopterus spinosus)
1RLW ☐	Lapwing (Vanellus vanellus)
2MW ☐	Knot (Calidris canutus)
1MW ☐	Sanderling (Calidris alba)
2MW ☐	Little Stint (Calidris minuta)
3M ☐	Temminck's Stint (Calidris temminckii)
1M ☐	Curlew Sandpiper (Calidris ferruginea)
4WL ☐	Purple Sandpiper (Calidris maritima)
1MW ☐	Dunlin (Calidris alpina)
1M ☐	Ruff (Philomachus pugnax)
4MW ☐	Jack Snipe (Lymnocryptes minimus)
1MW ☐	Snipe (Gallinago gallinago)
V ☐	Great Snipe (Gallinago media)
V ☐	Long-billed Dowitcher (Limnodromus scolopaceus)
V ☐	Short-billed Dowitcher (Limnodromus griseus)
4MWL ☐	Woodcock (Scolopax rusticola)
1MW ☐	Black-tailed Godwit (Limosa limosa)
2MW ☐	Bar-tailed Godwit (Limosa lapponica)
2M ☐	Whimbrel (Numenius phaeopus)
V ☐	Slender-billed Curlew (Numenius tenuirostris)
2MW ☐	Curlew (Numenius arquata)
2M ☐	Spotted Redshank (Tringa erythropus)
1RMW ☐	Redshank (Tringa totanus)
3M ☐	Marsh Sandpiper (Tringa stagnatilis)
2MW ☐	Greenshank (Tringa nebularia)
1MW ☐	Green Sandpiper (Tringa ochropus)
2M ☐	Wood Sandpiper (Tringa glareola)
1MW ☐	Common Sandpiper (Actitis hypoleucos)
V ☐	Spotted Sandpiper (Actitis macularia)
2MW ☐	Turnstone (Arenaria interpres)
V ☐	Wilson's Phalarope (Phalaropus tricolor)
5M ☐	Red-necked Phalarope (Phalaropus lobatus)
3M ☐	Grey Phalarope (Phalaropus fulicarius)
3M ☐	Pomarine Skua (Stercorarius pomarinus)
2MW ☐	Arctic Skua (Stercorarius parasiticus)
V ☐	Long-tailed Skua (Stercorarius longicaudus)
1MW ☐	Great Skua (Stercorarius skua)
1MW ☐	Mediterranean Gull (Larus melanocephalus)
V ☐	Franklin's Gull (Larus pipixcan)
V ☐	Laughing Gull (Larus atricilla)
2MW ☐	Little Gull (Larus minutus)
V ☐	Sabine's Gull (Larus sabini)
1MW ☐	Black-headed Gull (Larus ridibundus)
V ☐	Grey-headed Gull (Larus cirrocephalus)
4SL ☐	Slender-billed Gull (Larus genei)
1MW ☐	Audouin's Gull (Larus audouinii)
V ☐	Ring-billed Gull (Larus delawarensis)
3W ☐	Common Gull (Larus canus)
1MW ☐	Lesser Black-backed Gull (Larus fuscus)

1R ☐	Yellow-legged Herring Gull (Larus cachinnans)
V ☐	Iceland Gull (Larus glaucoides)
4W ☐	Great Black-backed Gull (Larus marinus)
2W ☐	Kittiwake (Rissa tridactyla)
2SML ☐	Gull-billed Tern (Gelochelidon nilotica)
2SMW ☐	Caspian Tern (Sterna caspia)
4M ☐	Royal Tern (Sterna maxima)
3M ☐	Lesser Crested Tern (Sterna bengalensis)
1MW ☐	Sandwich Tern (Sterna sandvicensis)
5M ☐	Roseate Tern (Sterna dougallii)
2SM ☐	Common Tern (Sterna hirundo)
5M ☐	Arctic Tern (Sterna paradisaea)
V ☐	Sooty Tern (Sterna fuscata)
1SML ☐	Little Tern (Sterna albifrons)
1SML ☐	Whiskered Tern (Chlidonias hybridus)
1SML ☐	Black Tern (Chlidonias niger)
V ☐	White-winged Black Tern (Chlidonias leucopterus)
3MW ☐	Guillemot (Uria aalge)
1MW ☐	Razorbill (Alca torda)
V ☐	Little Auk (Alle alle)
1M ☐	Puffin (Fratercula arctica)
2RL ☐	Black-bellied Sandgrouse (Pterocles orientalis)
2RL ☐	Pin-tailed Sandgrouse (Pterocles alchata)
4RL ☐	Rock Dove (Columba livia)
2WL ☐	Stock Dove (Columba oenas)
1R ☐	Woodpigeon (Columba palumbus)
V ☐	Collared Dove (Streptopelia decaocto)
1SM ☐	Turtle Dove (Streptopelia turtur)
2SML ☐	Great Spotted Cuckoo (Clamator glandarius)
1SM ☐	Cuckoo (Cuculus canorus)
1RL ☐	Barn Owl (Tyto alba)
1SM ☐	Scops Owl (Otus scops)
2RL ☐	Eagle Owl (Bubo bubo)
1R ☐	Little Owl (Athene noctua)
1R ☐	Tawny Owl (Strix aluco)
4RL ☐	Long-eared Owl (Asio otus)
3MW ☐	Short-eared Owl (Asio flammeus)
E ☐	African Marsh Owl (Asio capensis)
1SM ☐	Nightjar (Caprimulgus europaeus)
1SM ☐	Red-necked Nightjar (Caprimulgus ruficollis)
1SM ☐	Swift (Apus apus)
1SM ☐	Pallid Swift (Apus pallidus)
1SLM ☐	Alpine Swift (Apus melba)
2SL ☐	White-rumped Swift (Apus caffer)
V ☐	Little Swift (Apus affinis)
1RLW ☐	Kingfisher (Alcedo atthis)
V ☐	Blue-cheeked Bee-eater (Merops superciliosus)
1SM ☐	Bee-eater (Merops apiaster)
1SML ☐	Roller (Coracius garrulus)
1SLM ☐	Hoopoe (Upupa epops)

2MW ☐	Wryneck (Jynx torquilla)
2RL ☐	Green Woodpecker (Picus viridis)
1R ☐	Great Spotted Woodpecker (Dendrocopos major)
4RL ☐	Lesser Spotted Woodpecker (Dendrocopos minor)
4RL ☐	Dupont's Lark (Chersophilus duponti)
1RL ☐	Calandra Lark (Melanocorypha calandra)
1SML ☐	Short-toed Lark (Calandrella brachydactyla)
2SL ☐	Lesser Short-toed Lark (Calandrella rufescens)
1R ☐	Crested Lark (Galerida cristata)
1RL ☐	Thekla Lark (Galerida theklae)
1RM ☐	Woodlark (Lullula arborea)
4RL1MW ☐	Skylark (Alauda arvensis)
2M ☐	Sand Martin (Riparia riparia)
1RLW ☐	Crag Martin (Ptyonoprogne rupestris)
1SM ☐	Swallow (Hirundo rustica)
2SM ☐	Red-rumped Swallow (Hirundo daurica)
1SM ☐	House Martin (Delichon urbica)
V ☐	Richard's Pipit (Anthus novaeseelandiae)
1SLM ☐	Tawny Pipit (Anthus campestris)
1M ☐	Tree Pipit (Anthus trivialis)
1MW ☐	Meadow Pipit (Anthus pratensis)
4M ☐	Red-throated Pipit (Anthus cervinus)
4MWL ☐	Water Pipit (Anthus spinoletta)
4M ☐	Rock Pipit (Anthus petrosus)
1SM ☐	Yellow Wagtail (Motacilla flava)
1RW ☐	Grey Wagtail (Motacilla cinerea)
1W2RL ☐	White Wagtail (Motacilla alba)
V ☐	Common Bulbul (Pycnonotus barbatus)
4RL ☐	Dipper (Cinclus cinclus)
1R ☐	Wren (Troglodytes troglodytes)
2MWL ☐	Dunnock (Prunella modularis)
1RWL ☐	Alpine Accentor (Prunella collaris)
2SML ☐	Rufous Bush Robin (Cercotrichas galactotes)
1RMW ☐	Robin (Erithacus rubecula)
1SM ☐	Nightingale (Luscinia megarhynchos)
2MWL ☐	Bluethroat (Luscinia svecica)
1RLMW ☐	Black Redstart (Phoenicurus ochruros)
2SL1M ☐	Redstart (Phoenicurus phoenicurus)
1M ☐	Whinchat (Saxicola rubetra)
1RMW ☐	Stonechat (Saxicola torquata)
1SLM ☐	Wheatear (Oenanthe oenanthe)
1SM ☐	Black-eared Wheatear (Oenanthe hispanica)
V ☐	Desert Wheatear (Oenanthe deserti)
1RL ☐	Black Wheatear (Oenanthe leucura)
1SML ☐	Rock Thrush (Monticola saxatilis)
1RL ☐	Blue Rock Thrush (Monticola solitarius)
2MWL ☐	Ring Ouzel (Turdus torquatus)
1R ☐	Blackbird (Turdus merula)
3MW ☐	Fieldfare (Turdus pilaris)
1MW ☐	Song Thrush (Turdus philomelos)

2MW ☐	Redwing (Turdus iliacus)
1R ☐	Mistle Thrush (Turdus viscivorus)
1R ☐	Cetti's Warbler (Cettia cetti)
1R ☐	Fan-tailed Warbler (Cisticola juncidis)
2ML ☐	Grasshopper Warbler (Locustella naevia)
2SL ☐	Savi's Warbler (Locustella luscinioides)
4MW ☐	Moustached Warbler (Acrocephalus melanopogon)
V ☐	Aquatic Warbler (Acrocephalus paludicola)
2ML ☐	Sedge Warbler (Acrocephalus schoenobaenus)
V ☐	Blyth's Reed Warbler (Acrocephalus dumetorum)
5M ☐	Marsh Warbler (Acrocephalus palustris)
2SLM ☐	Reed Warbler (Acrocephalus scirpaceus)
1SM ☐	Great Reed Warbler (Acrocephalus arundinaceus)
2SML ☐	Olivaceous Warbler (Hippolais pallida)
V ☐	Icterine Warbler (Hippolais icterina)
1SM ☐	Melodious Warbler (Hippolais polyglotta)
V ☐	Marmora's Warbler (Sylvia sarda)
1RMW ☐	Dartford Warbler (Sylvia undata)
V ☐	Tristram's Warbler (Sylvia deserticola)
1SLM ☐	Spectacled Warbler (Sylvia conspicillata)
1SLM ☐	Subalpine Warbler (Sylvia cantillans)
1R ☐	Sardinian Warbler (Sylvia melanocephala)
2SLM ☐	Orphean Warbler (Sylvia hortensis)
V ☐	Lesser Whitethroat (Sylvia curruca)
1SLM ☐	Whitethroat (Sylvia communis)
1M ☐	Garden Warbler (Sylvia borin)
1RMW ☐	Blackcap (Sylvia atricapilla)
V ☐	Arctic Warbler (Phylloscopus borealis)
V ☐	Yellow-browed Warbler (Phylloscopus inornatus)
V ☐	Dusky Warbler (Phylloscopus fuscatus)
1SM ☐	Bonelli's Warbler (Phylloscopus bonelli)
3M ☐	Wood Warbler (Phylloscopus sibilatrix)
1RLMW ☐	Chiffchaff (Phylloscopus collybita)
1M ☐	Willow Warbler (Phylloscopus trochilus)
4MW ☐	Goldcrest (Regulus regulus)
1R ☐	Firecrest (Regulus ignicapillus)
1SM ☐	Spotted Flycatcher (Muscicapa striata)
V ☐	Red-breasted Flycatcher (Ficedula parva)
1M ☐	Pied Flycatcher (Ficedula hypoleuca)
2RL ☐	Long-tailed Tit (Aegithalos caudatus)
1R ☐	Crested Tit (Parus cristatus)
1RL ☐	Coal Tit (Parus ater)
1R ☐	Blue Tit (Parus caeruleus)
1R ☐	Great Tit (Parus major)
1RL ☐	Nuthatch (Sitta europaea)
4MW ☐	Wallcreeper (Tichodroma muraria)
1R ☐	Short-toed Treecreeper (Certhia brachydactyla)
1MWL ☐	Penduline Tit (Remiz pendulinus)
1SML ☐	Golden Oriole (Oriolus oriolus)
5M ☐	Red-backed Shrike (Lanius collurio)

V ☐	Lesser Grey Shrike (Lanius minor)
2RLMW ☐	Great Grey Shrike (Lanius excubitor)
1SM ☐	Woodchat Shrike (Lanius senator)
V ☐	Masked Shrike (Lanius nubicus)
1R ☐	Jay (Garrulus glandarius)
1RL ☐	Azure-winged Magpie (Cyanopica cyana)
1RL ☐	Magpie (Pica pica)
3? ☐	Alpine Chough (Pyrrhocorax graculus)
1R ☐	Chough (Pyrrhocorax pyrrhocorax)
1R ☐	Jackdaw (Corvus monedula)
5W ☐	Rook (Corvus frugilegus)
2RL ☐	Carrion Crow (Corvus corone)
2R ☐	Raven (Corvus corax)
1W ☐	Starling (Sturnus vulgaris)
1R ☐	Spotless Starling (Sturnus unicolor)
V ☐	Rose-coloured Starling (Sturnus roseus)
1R ☐	House Sparrow (Passer domesticus)
2RL ☐	Spanish Sparrow (Passer hispaniolensis)
2RL ☐	Tree Sparrow (Passer montanus)
2RL ☐	Rock Sparrow (Petronia petronia)
3RL ☐	Common Waxbill (Estrilda astrild)
1RMW ☐	Chaffinch (Fringilla coelebs)
3MW ☐	Brambling (Fringilla montifringilla)
1RMW ☐	Serin (Serinus serinus)
1RMW ☐	Greenfinch (Carduelis chloris)
1RMW ☐	Goldfinch (Carduelis carduelis)
1MW ☐	Siskin (Carduelis spinus)
1RMW ☐	Linnet (Acanthis cannabina)
2RL ☐	Crossbill (Loxia curvirostra)
4SML ☐	Trumpeter Finch (Bucanetes githagineus)
V ☐	Scarlet Rosefinch (Carpodacus erythrinus)
5M ☐	Bullfinch (Pyrrhula pyrrhula)
2RML ☐	Hawfinch (Coccothraustes coccothraustes)
V ☐	White-throated Sparrow (Zonotrichia albicollis)
V ☐	Slate-coloured Junco (Junco hyemalis)
V ☐	Snow Bunting (Plectrophenax nivalis)
V ☐	Pine Bunting (Emberiza leucocephalos)
5MW ☐	Yellowhammer (Emberiza citrinella)
2R ☐	Cirl Bunting (Emberiza cirlus)
1RL ☐	Rock Bunting (Emberiza cia)
V ☐	House Bunting (Emberiza striolata)
2SLM ☐	Ortolan Bunting (Emberiza hortulana)
V ☐	Little Bunting (Emberiza pusilla)
2MWL ☐	Reed Bunting (Emberiza schoeniclus)
V ☐	Black-headed Bunting (Emberiza melanocephala)
1RM ☐	Corn Bunting (Miliaria calandra)
V ☐	Bobolink (Dolichonyx oryzivorus)
☐	
☐	
☐	

MAMMALS

- ☐ Western Hedgehog (Erinaceus europaeaus)
- ☐ Algerian Hedgehog (Erinaceus algirus)
- ☐ Blind Mole (Talpa caeca)
- ☐ Pygmy White-toothed Shrew (Suncus etruscus)
- ☐ Greater White-toothed Shrew (Crocidura russula)
- ☐ Lesser White-toothed Shrew (Crocidura suaveolens)
- ☐ Lesser Horseshoe Bat (Rhinolophus hipposideros)
- ☐ Greater Horseshoe Bat (Rhinolophus ferrumequinum)
- ☐ Mediterranean Horseshoe Bat (Rhinolophus euryale)
- ☐ Daubenton's Bat (Myotis daubentoni)
- ☐ Natterer's Bat (Myotis nattereri)
- ☐ Greater Mouse-eared Bat (Myotis myotis)
- ☐ Lesser Mouse-eared Bat (Myotis blythi)
- ☐ Noctule (Nyctalus noctula)
- ☐ Serotine (Eptesicus serotinus)
- ☐ Common Pipistrelle (Pipistrellus pipistrellus)
- ☐ Kuhl's Pipistrelle (Pipistrellus kuhli)
- ☐ Savi's Pipistrelle (Pipistrellus savii)
- ☐ Grey Long-eared Bat (Plecotus austriacus)
- ☐ Schreiber's Bat (Miniopterus schreibersi)
- ☐ Free-tailed Bat (Tadarida teniotis)
- ☐ Rabbit (Oryctolagus cuniculus)
- ☐ Brown Hare (Lepus capensis)
- ☐ Garden Dormouse (Eliomys quercinus)
- ☐ Snow Vole (Microtus nivalis)
- ☐ Cabrera's Vole (Microtus cabrerae)
- ☐ Mediterranean Pine Vole (Pitymys duodecimcostatus)
- ☐ Southwestern Water Vole (Arvicola sapidus)
- ☐ Brown Rat (Rattus norvegicus)
- ☐ Black Rat (Rattus rattus)
- ☐ Wood Mouse (Apodemus sylvaticus)
- ☐ House Mouse (Mus musculus)
- ☐ Algerian Mouse (Mus spretus)
- ☐ Barbary Macaque (Macaca sylvanus)
- ☐ Wolf (Canis lupus)
- ☐ Red Fox (Vulpes vulpes)
- ☐ Weasel (Mustela nivalis)
- ☐ Western Polecat (Mustela putorius)
- ☐ Beech Marten (Martes foina)
- ☐ Badger (Meles meles)
- ☐ Otter (Lutra lutra)
- ☐ Egyptian Mongoose (Herpestes ichneumon)
- ☐ Genet (Genetta genetta)
- ☐ Lynx (Felis lynx)
- ☐ Wild Cat (Felis silvestris)
- ☐ Wild Boar (Sus scrofa)
- ☐ Mouflon (Ovis musimon)
- ☐ Spanish Ibex (Capra pyrenaicus)
- ☐ Red Deer (Cervus elaphus)

- ☐ Fallow Deer (Cervus dama)
- ☐ Roe Deer (Capreolus capreolus)
- ☐ Fin Whale (Balaenoptera physalus)
- ☐ Sperm Whale (Physeter catodon)
- ☐ Long-finned Pilot Whale (Globicephala melaena)
- ☐ False Killer Whale (Pseudorca crassidens)
- ☐ Killer Whale (Orcinus orca)
- ☐ Common Dolphin (Delphinus delphis)
- ☐ Striped Dolphin (Stenella coeruleoalba)
- ☐ Bottlenose Dolphin (Tursiops truncatus)

REPTILES AND AMPHIBIANS

- ☐ Spur-thighed Tortoise (Testudo graeca)
- ☐ European Pond Terrapin (Emys orbicularis)
- ☐ Stripe-necked Terrapin (Mauremys caspica)
- ☐ Moorish Gecko (Tarentola mauritanica)
- ☐ Turkish Gecko (Hemidactylus turcicus)
- ☐ Mediterranean Chameleon (Chamaeleo chamaeleon)
- ☐ Large Psammodromus (Psammodromus algirus)
- ☐ Spanish Psammodromus (Psammodromus hispanicus)
- ☐ Spiny-footed Lizard (Acanthodactylus erythrurus)
- ☐ Ocellated Lizard (Lacerta lepida)
- ☐ Iberian Wall Lizard (Podarcis hispanica)
- ☐ Italian Wall Lizard (Podarcis sicula)
- ☐ Bedriaga's Skink (Chalcides bedriagai)
- ☐ Three-toed Skink (Chalcides chalcides)
- ☐ Amphisbaenian (Blanus cinereus)
- ☐ Montpellier Snake (Malpolon monspessulanus)
- ☐ Horseshoe Whip Snake (Coluber hippocrepis)
- ☐ Ladder Snake (Elaphe scalaris)
- ☐ Grass Snake (Natrix natrix)
- ☐ Viperine Snake (Natrix maura)
- ☐ Southern Smooth Snake (Coronella girondica)
- ☐ False Smooth Snake (Macroprotodon cucullatus)
- ☐ Lataste's Viper (Vipera latasti)
- ☐ Fire Salamander (Salamandra salamandra)
- ☐ Sharp-ribbed Salamander (Pleurodeles waltl)
- ☐ Marbled Newt (Triturus marmoratus)
- ☐ Bosca's Newt (Triturus boscai)
- ☐ Painted Frog (Discoglossus pictus)
- ☐ Midwife Toad (Alytes obstetricans))
- ☐ Iberian Midwife Toad (Alytes cisternasii)
- ☐ Western Spadefoot (Pelobates cultripes)
- ☐ Parsley Frog (Pelodytes punctatus)
- ☐ Common Toad (Bufo bufo)

☐ Natterjack (Bufo calamita)
☐ Stripeless Tree Frog (Hyla meridionalis)
☐ Marsh Frog (Rana ridibunda)

BUTTERFLIES

☐ Swallowtail (Papilio machaon)
☐ Scarce Swallowtail (Iphiclides podalirius)
☐ Spanish Festoon (Zerynthia rumina)
☐ Apollo (Parnassius apollo)
☐ Large White (Pieris brassicae)
☐ Black-veined White (Aporia crataegi)
☐ Small White (Artogeia rapae)
☐ Green-veined White (Artogeia napi)
☐ Bath White (Pontia daplidice)
☐ Freyer's Dappled White (Euchloe simplonia)
☐ Portuguese Dappled White (Euchloe tagis)
☐ Green-striped White (Euchloe belemia)
☐ Orange Tip (Anthocharis cardamines)
☐ Moroccan Orange Tip (Anthocharis belia)
☐ Sooty Orange Tip (Zegris eupheme)
☐ Desert Orange Tip (Colotis evagore)
☐ Clouded Yellow (Colias crocea)
☐ Berger's Clouded Yellow (Colias australis)
☐ Brimstone (Gonepteryx rhamni)
☐ Cleopatra (Gonepteryx cleopatra)
☐ Wood White (Leptidea sinapis)
☐ Purple Hairstreak (Quercusia quercus)
☐ Spanish Purple Hairstreak (Laeosopis roboris)
☐ Ilex Hairstreak (Nordmannia ilicis)
☐ False Ilex Hairstreak (Nordmannia esculi)
☐ Blue-spot Hairstreak (Strymonidia spini)
☐ Green Hairstreak (Callophrys rubi)
☐ Provence Hairstreak (Tomares ballus)
☐ Small Copper (Lycaena phlaeas)
☐ Purple-shot Copper (Heodes alciphron)
☐ Long-tailed Blue (Lampides boeticus)
☐ Lang's Short-tailed Blue (Syntarucus pirithous)
☐ African Grass Blue (Zizeeria knysna)
☐ Little Blue (Cupido minimus)
☐ Osiris Blue (Cupido osiris)
☐ Lorquin's Blue (Cupido lorquinii)
☐ Holly Blue (Celastrina argiolus)
☐ Green-underside Blue (Glaucopsyche alexis)
☐ Black-eyed Blue (Glaucopsyche melanops)
☐ Iolas Blue (Iolana iolas)
☐ Panoptes Blue (Pseudophilotes panoptes)
☐ False Baton Blue (Pseudophilotes abencerragus)

- ☐ Chequered Blue (Scolitantides orion)
- ☐ Zephyr Blue (Plebejus pylaon)
- ☐ Silver-studded Blue (Plebejus argus)
- ☐ Idas Blue (Lycaeides idas)
- ☐ Brown Argus (Aricia agestis)
- ☐ Mountain Argus (Aricia artaxerxes)
- ☐ Spanish Argus (Aricia morronensis)
- ☐ Glandon Blue (Agriades glandon)
- ☐ Mazarine Blue (Cyaniris semiargus)
- ☐ Chapman's Blue (Agrodiaetus thersites)
- ☐ Escher's Blue (Agrodiaetus escheri)
- ☐ Amanda's Blue (Agrodiaetus amanda)
- ☐ Mother-of-pearl Blue (Plebicula nivescens)
- ☐ Spanish Chalk-hill Blue (Lysandra albicans)
- ☐ Adonis Blue (Lysandra bellargus)
- ☐ Common Blue (Polyommatus icarus)
- ☐ Nettle-tree Butterfly (Libythea celtis)
- ☐ Two-tailed Pasha (Charaxes jasius)
- ☐ Large Tortoiseshell (Nymphalis polychloros)
- ☐ Red Admiral (Vanessa atalanta)
- ☐ Painted lady (Cynthia cardui)
- ☐ Small Tortoiseshell (Aglais urticae)
- ☐ Comma Butterfly (Polygonia c-album)
- ☐ Cardinal (Pandoriana pandora)
- ☐ Silver-washed Fritillary (Argynnis paphia)
- ☐ Dark Green Fritillary (Mesoacidalia aglaja)
- ☐ High Brown Fritillary (Fabriciana adippe)
- ☐ Niobe Fritillary (Fabriciana niobe)
- ☐ Queen of Spain Fritillary (Issoria lathonia)
- ☐ Twin-spot Fritillary (Brenthis hecate)
- ☐ Knapweed Fritillary (Melitaea phoebe)
- ☐ Aetherie Fritillary (Melitaea aetherie)
- ☐ Spotted Fritillary (Melitaea didyma)
- ☐ Lesser Spotted Fritillary (Melitaea trivia)
- ☐ Heath Fritillary (Mellicta athalia)
- ☐ Provençal Fritillary (Mellicta deione)
- ☐ Meadow Fritillary (Mellicta parthenoides)
- ☐ Marsh Fritillary (Eurodryas aurinia)
- ☐ Spanish Fritillary (Eurodryas desfontainii)
- ☐ Marbled White (Melanargia galathea)
- ☐ Western Marbled White (Melanargia occitanica)
- ☐ Spanish Marbled White (Melanargia ines)
- ☐ Rock Grayling (Hipparchia alcyone)
- ☐ Common Grayling (Hipparchia semele)
- ☐ Tree Grayling (Neohipparchia statilinus)
- ☐ Striped Grayling (Pseudotergumia fidia)
- ☐ Hermit (Chazara briseis)
- ☐ Black Satyr (Satyrus actaea)
- ☐ Great Banded Grayling (Brintesia circe)
- ☐ False Grayling (Arethusana arethusa)

- [ ] Spanish Brassy Ringlet (Erebia hispania)
- [ ] Meadow Brown (Maniola jurtina)
- [ ] Dusky Meadow Brown (Hyponephele lycaon)
- [ ] Oriental Meadow Brown (Hyponephele lupina)
- [ ] Ringlet (Aphantopus hyperantus)
- [ ] Gatekeeper (Pyronia tithonus)
- [ ] Southern Gatekeeper (Pyronia cecilia)
- [ ] Spanish Gatekeeper (Pyronia bathseba)
- [ ] Small Heath (Coenonympha pamphilus)
- [ ] Dusky Heath (Coenonympha dorus)
- [ ] Speckled Wood (Pararge aegeria)
- [ ] Wall Brown (Lasiommata megera)
- [ ] Large Wall Brown (Lasiommata maera)
- [ ] Monarch (Danaus plexippus)
- [ ] Grizzled Skipper (Pyrgus malvae)
- [ ] Large Grizzled Skipper (Pyrgus alveus)
- [ ] Oberthur's Grizzled Skipper (Pyrgus armoricanus)
- [ ] Olive Skipper (Pyrgus serratulae)
- [ ] Carline Skipper (Pyrgus carlinae)
- [ ] Rosy Grizzled Skipper (Pyrgus onopordi)
- [ ] Safflower Skipper (Pyrgus carthami)
- [ ] Red-underwing Skipper (Spialia sertorius)
- [ ] Sage Skipper (Syrichtus proto)
- [ ] Mallow Skipper (Carcharodus alceae)
- [ ] Marbled Skipper (Carcharodus lavatherae)
- [ ] Southern Marbled Skipper (Carcharodus boeticus)
- [ ] Tufted Marbled Skipper (Carcharodus flocciferus)
- [ ] Dingy Skipper (Erynnis tages)
- [ ] Lulworth Skipper (Thymelicus acteon)
- [ ] Essex Skipper (Thymelicus lineolus)
- [ ] Small Skipper (Thymelicus flavus)
- [ ] Silver-spotted Skipper (Hesperia comma)
- [ ] Mediterranean Skipper (Gegenes nostrodamus)
- [ ] Zeller's Skipper (Borbo borbonica)

DRAGONFLIES

- [ ] Mediterranean Demoiselle (Calopteryx haemorrhoidalis)
- [ ] Beautiful Demoiselle (Calopteryx virgo)
- [ ] (Calopteryx xanthostoma)
- [ ] Common Winter Damselfly (Sympecma fusca)
- [ ] Southern Emerald Damselfly (Lestes barbarus)
- [ ] Small Emerald Damselfly (Lestes virens)
- [ ] Willow Emerald Damselfly (Lestes viridis)
- [ ] Dark Emerald Damselfly (Lestes macrostigma)
- [ ] Emerald Damselfly (Lestes sponsa)

☐ Scarce Emerald Damselfly (Lestes dryas)
☐   (Platycnemis acutipennis)
☐   (Platycnemis latipes)
☐ Large Red Damselfly (Pyrrhosoma nymphula)
☐ Scarce Blue-tailed Damselfly (Ischnura pumilio)
☐   (Ischnura graellsii)
☐ Goblet-marked Damselfly (Coenagrion lindenii)
☐ Dainty Damselfly (Coenagrion scitulum)
☐ Southern Damselfly (Coenagrion mercuriale)
☐   (Coenagrion caerulescens)
☐ Azure Damselfly (Coenagrion puella)
☐ Common Blue Damselfly (Enallagma cyathigerum)
☐ Small Red-eyed Damselfly (Erythromma viridulum)
☐ Small Red Damselfly (Ceriagrion tenellum)
☐ Western Club-tailed Dragonfly (Gomphus pulchellus)
☐ Yellow Club-tailed Dragonfly (Gomphus simillimus)
☐   (Gomphus graslini)
☐   (Paragomphus genei)
☐ Green-eyed Hook-tailed Dragonfly (Onychogomphus forcipatus)
☐ Blue-eyed Hook-tailed Dragonfly (Onychogomphus uncatus)
☐   (Onychogomphus costae)
☐ Crepuscular Hawker (Boyeria irene)
☐ Southern Hawker (Aeshna cyanea)
☐ Migrant Hawker (Aeshna mixta)
☐ Mediterranean Hawker (Aeshna affinis)
☐ Norfolk Hawker (Anaciaeschna isosceles)
☐ Vagrant Emperor Dragonfly (Hemianax ephippiger)
☐ Emperor Dragonfly (Anax imperator)
☐ Lesser Emperor Dragonfly (Anax parthenope)
☐ Golden-ringed Dragonfly (Cordulegaster boltonii)
☐ Broad-bodied Chaser (Libellula depressa)
☐ Scarce Chaser (Libellula fulva)
☐ Four-spotted Chaser (Libellula quadrimaculata)
☐ Black-tailed Skimmer (Orthetrum cancellatum)
☐ Keeled Skimmer (Orthetrum coerulescens)
☐ Southern Skimmer (Orthetrum brunneum)
☐   (Orthetrum trinacria)
☐   (Orthetrum chrysostigma)
☐   (Orthetrum nitidinerve)
☐   (Diplacodes lefebvrii)
☐ Scarlet Darter (Crocothemis erythraea)
☐   (Brachythemis leucosticta)
☐ Ruddy Darter (Sympetrum sanguineum)
☐ Yellow-winged Darter (Sympetrum flaveolum)
☐ Red-veined Darter (Sympetrum fonscolombii)
☐ Southern Darter (Sympetrum meridionale)
☐ Common Darter (Sympetrum striolatum)
☐   (Zygonyx torrida)
☐   (Trithemis annulata)
☐
☐

SOCIETIES AND CLUBS

THE GIBRALTAR ORNITHOLOGICAL AND NATURAL HISTORY SOCIETY

The object of the Gibraltar Ornithological and Natural History Society (a National Section of BirdLife International) is the study and conservation of the wildlife of the area of the Strait of Gibraltar.

The Society:
- Runs a volunteer Strait of Gibraltar Bird Observatory
- Publishes an annual journal 'Alectoris' which includes the annual report of the Strait Observatory
- Publishes quarterly newsletters for members
- Publishes quarterly reports with tables, summarising the Strait Observatory's results
- Is responsible for the Gibraltar Rarities Committee
- Holds regular meetings, with lectures, and outings for its members

Visitors to the area of the Strait are encouraged to submit their observations to the Records Officer, Gibraltar Ornithological & Natural History Society, PO Box 843, Gibraltar.

SOCIEDAD ESPANOLA DE ORNITOLOGIA

This is the official ornithological body in Spain. It co-ordinates records and ringing of birds in Spain . The Sociedad publishes the scientific journal 'Ardeola' and a bulletin 'La Garcilla' which includes information and papers on southern Spain. The Sociedad has an Iberian Rarities Panel and observations of rare birds seen in Spain should be sent to the Secretario, Comite Iberico de Rarezas, Sociedad Española de Ornitologia, Catedra de Zoologia (Vertebrados), Facultad de Biología, Universidad Complutense, 28040 Madrid.

AGENCIA DEL MEDIO AMBIENTE

The official Andalucian Government environment agency is the Agencia del Medio Ambiente (AMA). AMA has a network of wardens which look after protected sites (eg Cádiz lagoons, Grazalema Fir wood, etc). The AMA has offices in each of the provincial capitals and its headquarters are in Seville. The address is Agencia del Medio Ambiente, Avenida de Eritaña, 2, 41013 - Sevilla.

BIBLIOGRAPHY

Arnold, E.N. & Burton, J.A (1978) A Field Guide to the Reptiles and Amphibians of Britain and Europe. Collins.

Bernis, F. (1980) La Migracion de las aves en el estrecho de Gibraltar: Vol 1. Madrid University.

Corbet, G. & Ovenden, D. (1980) The Mammals of Britain and Europe. Collins.

Cortes, J.E. et al. (1980) The Birds of Gibraltar. Gibraltar Bookshop.

Cortes, J.E. & Finlayson, J.C. (1988) The Flowers and Wildlife of Gibraltar. Gibraltar Bookshop.

D'Aguilar, J. et al. (1986) A Field Guide to the Dragonflies of Britain, Europe and North Africa. Collins.

Finlayson, J.C. (1992) Birds of the Strait of Gibraltar. Poyser.

Finlayson, J.C. & Cortes, J.E. (1987) The Birds of the Strait of Gibraltar - its waters and northern shore. Alectoris 6, Gibraltar Ornithological Society Report.

Grant, P.J. (1982) Gulls - a guide to identification. Poyser.

Grimmett, R.F.A. & Jones, T.A. (1989) Important Bird Areas in Europe. ICBP.

Grunfeld, F.V. (1988) Wild Spain. Edbury Press.

Higgins, L. & Hargreaves, B. (1983) The Butterflies of Britain and Europe. Collins.

Humphries, C.J. et al. (1985) Trees of Britain and Europe. Country Life Guides.

Jackson, Sir W.F.G. (1987) The Rock of the Gibraltarians - a history of Gibraltar. Fairleigh Dickinson University Press.

Luard, N. (1984) Andalucía - A Portrait of Southern Spain. Century.

Petersen, R.T. et al. A Field Guide to the Birds of Britain and Europe. Collins.

Polunin, O. & Huxley, A. (1974) Flowers of the Mediterranean. Chatto & Windus.

Polunin, O. & Smythies, B.E. (1988) Flowers of south-west Europe - a Field Guide. Oxford University Press.

Porter, R.F. et al. (1981) Flight Identification of European Raptors. Poyser.

Telleria, J.L. (1981) La Migracion de las aves en el estrecho de Gibraltar: Vol 2. Madrid University.